Peter Sprent was born in Australia in 1923 and is now an adopted Scot who has lived in the ancient Kingdom of Fife since 1967. He is Emeritus Professor of Statistics at the University of Dundee and sees retirement as the time to do the things one really likes. Stimulated by experience as a consultant in fields ranging from agriculture and biology to industry and commerce, his mission is to take the mystery out of mathematics and the stuffiness out of statistics. He is the author of *Statistics in Action*, *Quick Statistics* and *Understanding Data*, all published by Penguin, and has written elementary and advanced texts and numerous papers for professional journals.

Peter Sprent is a Fellow of the Royal Society of Edinburgh, a member of the International Statistical Institute and other professional bodies. His hobbies include hillwalking, flying and golf, but he confesses his scores at that game would bring joy to a first-class cricketer.

PETER SPRENT

# TAKING RISKS
## THE SCIENCE OF UNCERTAINTY

PENGUIN BOOKS

PENGUIN BOOKS

Published by the Penguin Group
27 Wrights Lane, London w8 5tz, England
Viking Penguin Inc., 40 West 23rd Street, New York, New York 10010, USA
Penguin Books Australia Ltd, Ringwood, Victoria, Australia
Penguin Books Canada Ltd, 2801 John Street, Markham, Ontario, Canada l3r 1b4
Penguin Books (NZ) Ltd, 182–190, Wairau Road, Auckland 10, New Zealand

Penguin Books Ltd, Registered Offices: Harmondsworth, Middlesex, England

First published 1988

Made and printed in Great Britain by
Richard Clay Ltd, Bungay, Suffolk
Filmset in Monophoto Photina

# CONTENTS

# PREFACE

Whims of nature, man's follies as well as his quest for material wealth and comfort, and new technologies all fuel the fires of uncertainty.

Perversely, since the beginning of time, man has made only minimal effort to understand the dominant role that uncertainty plays in determining his fate. He has sought relief from its more unpleasant manifestations in religion and prayer. Or turned a blind eye, adopting a fatalistic stance: accepting its benefits as strokes of good fortune.

Ignorance about risks is widespread. It is easy to sweep uncertainty under the carpet, to dismiss its more daunting manifestations as acts of God, or the devil; yet a better understanding is important to us all, especially to administrators, politicians, managers, engineers, technologists, journalists and broadcasters, and also to experts in, and students of, the sciences and social sciences. By appreciating the nature, the quirks, of uncertainty, we may take better account of it in decision making. There are no easy cure-alls for the problems it brings, but understanding equips us to cope with them in a way that will make life pleasanter for us and our descendants.

A serious scientific study of uncertainty began some 400 years ago when mathematicians used their skills to advise gamblers, but only in this century have we developed techniques to assess those manifestations that pervade science, engineering, industry, commerce, our health, our social system, indeed our very lifestyle. Understanding becomes ever more important as modern technology brings new benefits together with new risks, increasing the uncertainty of an inherently uncertain world.

In this book we look at some of the scientific tools for dealing with uncertainty and soon learn that these cannot solve all the problems, since there are important psychological and social elements that we ignore at our peril.

The two parts of the book differ in emphasis. Part I is concerned with techniques for, and difficulties in, a quantitative study of uncertainty. Some, but not all, of this material has to do with probability theory and its statistical applications. In Part II we look at real-world examples to highlight the problems, and describe rational ways to cope with them. We examine first approaches to major problems that have both national and international dimensions. Less spectacular, but more familiar, uncertainties are met from Chapter 10 onward.

The book is a series of connected essays, with emphasis on applied common sense rather than mathematical mystique. Where we look at problems that have been solved, it will be clear this has happened because people have used all available information in a rational way. Ignoring relevant information – or using only those bits that suit the arguments of interested parties – is a dangerous and disturbing tendency.

My aim is to encourage decision makers, and those affected by the decisions, to examine all arguments critically and to evaluate policies where risks or uncertainties are involved by rational assessment of the best available information.

Sources of illustrative data are acknowledged at the foot of relevant tables. I am indebted to the Royal Society, the Scottish Wildlife Trust, the Social and Economic Research Council, the Royal Statistical Society and Dr D. R. Green for permission to quote copyright material acknowledged in the text.

My thanks are due to Peter Walford for his patient and helpful guidance during my visit to the Cape Grim Baseline Air Pollution Monitoring Station.

P. Sprent
January 1988

# PART I

UNCERTAINTY: THE TECHNICAL
BACKGROUND

# CHAPTER I

## DOUBT AND REALITY

### Statesmen and politicians

On 13 November 1789, Benjamin Franklin, statesman and scientist, wrote to Jean Baptiste Le Roy:

> But in this world nothing can be said to be certain, except death and taxes.

Cynical, pessimistic and not completely true. A few primitive societies pay no taxes. Death is sure, but its timing unpredictable. Many events are, *for all practical purposes*, certain: be it in London, Sydney or Timbuktu, barring some remote catastrophe, the sun will rise tomorrow at a time predicted by astronomers of old.

Yet Franklin was right to imply uncertainty the rule rather than the exception. In our fast-changing twentieth century, a better understanding of the nature of uncertainty, of the reality of the problems it brings, has a new importance; an informed approach is essential for wise decision making. Collecting and assessing relevant information is important, but all too often urgently needed action is postponed with the politically cosy advice 'wait until we are sure'. This is often an excuse to delay an awkward decision on anything from building a nuclear power plant or banning the use of dangerous chemicals to framing laws to reduce carnage on the roads, and exemplifies an attitude that exposes widespread ignorance about the nature of uncertainty and procrastination that savours more of political opportunism than statesmanship. For, about many things, we shall never be sure: we must balance the risks against the benefits.

## A plea

> In a world of uncertainty it is time that we began to encourage our
> pupils to talk about and explore the uncertain rather than avoid it
> and hope it will go away.

This was how Green (1987) expressed the need for an informed
approach to uncertainty in schools. His plea could be addressed to
politicians, administrators, professional and business leaders, and
to some scientists who add unthinkingly to today's uncertainties
or play no constructive part in resolving them.

Ignorance about uncertainty is also a media weakness,
common among news editors, commentators, run-of-the-mill journalists, who, loath as we may be to admit it, all too often shape
our views. Vociferous pressure groups, some displaying an almost
fiendish delight in playing on public fear of risk, could make a
greater contribution to a secure future if they showed more understanding of the real nature of uncertainty, sought out all relevant
information about it and examined it rationally, rather than selecting only those bits that bolster their case.

Widespread misconceptions about risk cause daily frustration,
may lower industrial efficiency and may even create a serious
threat to human well-being.

## A rational approach

A scientific study of uncertainty began when seventeenth century
gamblers sought the advice of eminent mathematicians about
betting strategies, a frivolous and rather late start for the proper
study of a phenomenon that has been with us from the beginning
of time. Nature is a master of catastrophic uncertainty: the great
flood was an early reminder. Even if one does not accept the
veracity of the biblical description, there is evidence from other
cultures that floods were an early cause of devastation. The
science of probability and statistics, while dealing adequately with
many everyday problems of uncertainty as well as helping (or
warning) gamblers, is still of limited (but slowly increasing) use
for predicting floods or earthquakes.

A serious study of risk and uncertainty must take account of more than formal statistical methods. It calls for an understanding of how to collect and interpret relevant factual evidence. Equally importantly, it demands an appreciation of the psychology of human reaction to uncertainty and risk; this is linked both to how we perceive danger and to our ability to interpret correctly *all* relevant factual information.

This book uses examples to explain some of the problems of deep-rooted uncertainties and what has been, or might be, done about them. It sketches the path to rational understanding, highlights difficulties and stresses the need for informed compromise.

It does not provide solutions for complex and controversial problems. It makes a plea for rational decision making in the realization that uncertainty and risk always have been, and always will be, with us.

## Not all gloom

The story of uncertainty is not one of unmitigated gloom. Man has turned its challenges to advantage, harnessing the laws of chance. In industry, statistical quality control, developed by British and American workers, applied enthusiastically by the Japanese, has brought us more reliable goods at lower prices. Statistical pattern recognition has enabled us to squeeze information from fuzzy satellite photographs of immense value to agriculturists, foresters, meteorologists and others; the same techniques applied in medicine have eased problems of diagnosis, ensuring quicker and more reliable detection of diseases where early treatment is essential. In everyday life uncertainty may bring unanticipated comforts, ranging from a £1 000 000 win on the football pools to an unexpected gift or a chance meeting with somebody who will greatly influence our future.

Gloomy manifestations of uncertainty arouse fears of nuclear war and focus on the less often discussed, but perhaps greater, dangers of chemical or biological warfare and the possibility of air disasters – risks we would prefer to be without; dangers all rational people want to minimize. To individuals it may bring niggling minor irritations or be a harbinger of near or distant disaster; Bhopal and Chernobyl are synonymous with unexpected, un-predicted (except with hindsight) accidents, their consequences far reaching. Major mishaps, often striking without warning, are a price we pay for the benefits of high technology. Is that price too high?

How then do we cope rationally with uncertainty? What is the role of science? Can it provide *all* the answers? The second question is quickly disposed of with one word – 'no'. But let us not be too depressed, for logic and science are of considerable help.

### The quantitative element: probability

The cornerstone for the scientific approach is *probability and statistics*, ideal when we have quantitative data and a mathematical model of a physical situation. We may assert with confidence that the probability a heavy cigarette smoker will contract lung cancer is about twenty-four times as great as that facing a non-smoker; also that a heavy smoker is one-and-a-half times as likely as a non-smoker to die from cardiac or circulatory disease. Even solid information like this is easily misinterpreted. Many people think this means that those who smoke heavily are more likely to die of lung cancer than cardiac or circulatory disease. That is not so: they are more likely to die of the latter, since in the community at large it is a much commoner cause of death than lung cancer. This common misunderstanding shows that care is needed in making deductions from probability-type statements.

Here is another example where we use probability arguments. In most industrial processes, there is the bugbear of faulty items being produced in a more or less unpredictable way (*at random* is the technical jargon). We want to keep that number as low as possible without making costs too high. Manufacturers check whether there are too many faulty items by taking a sample of, say, a hundred items from a production run. Suppose we find two of these are defective; probability theory is then able to tell us we can be 95% certain that there will not be more than 4.3% defective in the whole production run (which may consist of thousands, perhaps even millions, of items). This is important information for a manufacturer if he guarantees not more than 5% defective items in batches he delivers to a customer. If he were guaranteeing not more than 1% defective, he would need to look whether something had gone wrong if he got two defective items (i.e. 2%) in a sample of a hundred.

## Casinos and bookmakers

Gambling casinos make money from roulette, blackjack and other games because the laws of probability ensure for them a long-term expected gain. The bookmaker is skilled in choosing the odds he offers on sporting events (and less sporting events like parliamentary elections), and in adjusting these in accord with amounts invested, in a way that ensures he too has a long-term expected gain. By and large most who bet with these professionals must lose, although a few regular punters with special knowledge and skills may consistently win, leaving the bookmaker to reap his profit from the greater number of speculators who rely on newspaper tips or the expedient of sticking a pin in a list of runners, or even backing a horse called Matilda because their Great-aunt Matilda is so nice.

## Utility

Betting illustrates another aspect of behaviour under uncertainty: while the casino operator or the skilled bookmaker is guaranteed long-term gains by the laws of probability, the rational gambler knows this means that in most circumstances he will, in the long run, lose. But is the homebound office clerk who wagers a pound on the favourite for tomorrow's 3.30 being any more foolish than he is in insuring his belongings against fire and theft? In the long run the insurance company makes a profit; the wise customer wants it to do so (in moderation), so that it has resources to pay any claim he might make. This means few of us in our lifetime collect more in insurance claims than we pay in premiums. Yet even those who find it irrational to bet usually insure their property. Their behaviour is explained by an extension of probabilistic and statistical notions to what is called *utility*.

Insurance against misfortune is common. But a shrewd Japanese golfer insures against good luck, the possibility of a hole in one; for tradition there demands that he who so scores must treat his partners and his and their caddies to dinner, offer souvenirs to his fellow club members and plant a commemorative tree, all of which is likely to cost him a goodly number of yen – the equivalent of well over £1000.

Unlike classical probability theory – with roots in mathematical rules that do not bow to the will of man – utility has a personal element. This is a complication in coping with uncertainty, for utility for a particular action differs from person to person, from community to community. To a large group of jobless in a depressed area, an announcement that a nuclear power station is to be built nearby may be greeted with enthusiasm: they assign high utility to the project. It will meet with less enthusiasm from workers in a coal mine twenty miles away who learn that the new plant is to replace the local coal-fired power station, or from the regional anti-nuclear committee: both groups assign low utilities, albeit for different reasons.

Given adequate information, we may quantify utilities (something we look at in Chapter 4), but the problems of uncertainty do not end here.

## Personal assessments of risk

We have a lot of data (factual information) for assessing relative safety of different modes of travel, yet many people do not use that evidence to make safety-oriented decisions.

One is more likely to be killed travelling $x$ miles in a motor car than if one covers the same distance in an airliner. Yet, given both options, the number who prefer the highway because they think flying is 'not really safe' exceeds those who take the opposite view and fly whenever possible because road travel is less safe; personal assessment of danger often bears little resemblance to a logical ordering of risks. Of course, many people do know road travel is less safe, but even so regard the danger as still small and make their choice on other grounds – factors such as costs, convenience, etc. These people assign a higher *utility* to road travel than they do to flying.

It is harder to be rational about different risks when there is a lack of relevant information, or if we have to make complicated calculations or deductions with the danger that they may be carried out incorrectly. Lack of information can often be overcome, but it may be costly.

Currently (in 1988) there is serious (and justifiable) alarm about a health threat (as well as one to climate) from a slow depletion of ozone in the upper atmosphere. Scientists worldwide are busy looking for an explanation and theories abound. The matter is urgent, and action is needed to reverse the trend if it is indeed man-made; we need more information. We take a detailed look at this problem in Chapter 9.

Even when we have relevant data, erroneous interpretations are all too easily made and can lead people astray, a problem already illustrated by the common belief that heavy smokers are more likely to die from lung cancer than cardiac problems. We look at other examples in Chapter 3.

## Catastrophes

The very unlikely one-off event that has devastating consequences when it occurs represents another kind of uncertainty, largely outside the conventional probability framework.

Floods, earthquakes, tidal waves, typhoons and so on may take man by surprise (although we are gradually improving our forecasting techniques), but science and technology have helped to reduce their dangers, to mitigate the consequences. At the same time modern technology has produced new potential sources of catastrophe. Where will tomorrow's Bhopals and Chernobyls occur? That there will be such disasters, and worse, few of us doubt. We are often not well equipped to make rational assessments of how far we can or should reduce such risks, but that is no excuse to put off decisions to that never-to-be day 'when we are sure'.

*Fears, justifiable or not?*

A much publicized example of what many see as a potential source of catastrophe has haunted Britain for several years. How should we dispose of nuclear waste? There is near unanimity from every sizeable community, even from remote settlements, that it should be buried 'somewhere else'. Such was the relief, on 1 May 1987, when plans to dump nuclear waste in their neighbourhood were abandoned, that villagers of Elstow in Bedfordshire celebrated by drinking champagne on the test-site.

There is a wealth of evidence that, with proper packaging and due regard to geological structure, this waste may be disposed of in a way that makes it less dangerous than natural radioactive deposits. Yet some objectors openly take the line that it must stay at the plants producing it until we are 'sure' burial is safe. The management of such plants doubtless take reasonable care and, while transport from sites must be carefully controlled, the material is certainly less safe at source of production than if properly buried. At plants it is more exposed to accidents that might result in its dispersal, to sabotage, to enemy or terrorist attack, all with potentially serious if not disastrous consequences.

Radiation from a properly sited nuclear waste dump is unlikely to exceed 1% of that from natural sources. Indeed, in parts of Britain, people are exposed daily to the gas radon, a product of the natural decay of uranium in the earth's crust, giving radiation levels 300 times that likely from a nuclear waste dump. Radon gets scant publicity from the anti-nuclear lobby; it might divert attention from the campaign against nuclear power. The nuclear industry has only recently thrown off a reluctance to mention it – presumably having thought any publicity for radiation was bad.

Scientists responsible for solving this waste problem are not immune from blame for the hysteria it has generated. They have clearly failed to get their safety message over to the public, or even to the politicians and to the more responsible media elements who inevitably influence, if not shape, public opinion in these matters.

## Looking ahead

In the remaining chapters of Part'I, we discuss people's perception of, and attitudes to, risk; describe common misconceptions and mistakes in data handling; then look at probability, utility and quantitative assessment of risk, developing arguments by simple examples rather than from formal theory.

In the rest of this chapter, we look briefly at a variety of situations where uncertainty plays a key role, thereby giving an idea of its complex nature. Some of these topics are expanded in the case histories in Part II.

## Everyday uncertainty

Our daily routine is riddled with uncertainty; it is usually all pervasive but undramatic. Our train or bus might be on time, or five, six or even ten minutes late; our work frustrating, dreary tasks taking longer than expected, a potential client failing to turn up, the office telephone ringing incessantly, our computer breaking down; the queue in the bank moving all too slowly; the bar lunch in that favourite pub not up to standard.

Every day we solve, often subconsciously, risk–benefit equations about routine matters. Each time we cross a road we *might* be run over; yet crossing the road is important, perhaps essential, to our routine. A rational person minimizes the risk by using a pedestrian crossing, a subway, or waiting for a break in the traffic stream. *The Highway Code* is a manual on risk reduction.

A spouse's greeting of 'Have a good day at the office, dear?' is a way of asking 'How did the fates (or uncertainties) treat you today?'

That weekly flutter on the football pools or a snap decision to buy shares in ICI or British Gas or BP carry risks and benefits of differing magnitudes; risks that are greater than putting one's money in a bank or building society, but so are the potential benefits. Whether somebody puts a £1000 windfall into a building society, into blue-chip shares or into Premium Bonds tells us a lot about their attitude to risks. Personal attitudes are not immutable:

they may be changed by one dramatic, unpredicted event. The Wall Street crash of 1929 and the Stock Market crisis in October 1987 both changed many people's attitudes to investment risks; Chernobyl, their assessment of the risks from nuclear power.

### Safety hazards

We may take everyday minor risks for granted, but the human lifespan is determined by interacting factors, a balancing of benefits and risks. Advances in medical science hold potential for healthier, longer lives, while industrialization has unleashed new dangers from hazardous by-products and environmental pollution.

More research has been done on that hot potato, nuclear waste, than on the safe dumping of toxic but *non*-radioactive chemicals. Unlike nuclear waste, dumping of toxic substances has already led to deaths and mass evacuations. Science writer James Bellini pulled no punches in describing dangers from this and other indus-trial activities in *High Tech Holocaust* (1986). The relation between size of potential catastrophes, however remote, and the risk we associate with them has been studied by many psychologists. This is something we look at in Chapter 2.

### *Our health*

New threats to health are regularly unveiled; some are associated with industry, others with the food we eat. We have long known that asbestos dust carries a risk of lung cancer – a risk multiplied manyfold if one is also a heavy smoker. American data indicate that, before steps were taken to avoid inhalation of asbestos dust by workers, among those who were non-smokers the incidence rate for lung cancer was five times that for other non-smokers, and fifty times higher for an asbestos worker who smoked.

Fatty foods may clog the arteries of those whose metabolism cannot cope with cholesterol, but not all of us have that difficulty. Some colouring or flavouring additives once used freely have been banned because we now know more about carcinogens. Evidence is mounting that many allergies, especially in children, are attributable to food additives; some are there to make the food more palatable, others just to prolong shelf life. We need more evidence to pinpoint the culprits.

## World threats

Many of today's threats have a global dimension. That industrial pollution is harming the environment is not disputed. But how badly? A threat to life on earth, say some. Total disaster is not yet imminent, but there is a pressing need to weigh carefully relevant evidence, to make the best logical assessment. These are matters we look at in Chapters 8 and 9.

Does the undeniable possibility of a nuclear holocaust, or a catastrophic radiation leak, justify scrapping all civilian atomic energy projects? Is this the simple way to avoid all possibility of a second, perhaps worse, Chernobyl? If you answer yes, pause a moment. Might not abandoning peaceful nuclear projects increase the danger from accidents at 'secret' military developments? For defence and security reasons, safety at these may not receive the priority it would in planning and operating a public utility subject to regulations and external scrutiny. On 1 January 1988, government documents released under the thirty year rule revealed that the published report on the fire in October 1957 at the Windscale nuclear reactor in Cumberland had been what might be described in topical idiom as 'economical with the truth'. Of course, if the major world powers ban nuclear weapons (and successfully police that ban) the argument above would lose much of its force.

### Killers, old and new

Medical science has eliminated many killer diseases. Tuberculosis, poliomyelitis and smallpox have been almost completely eradicated in many parts of the world; sadly there are pockets of resistance in less affluent countries. At the same time changes in attitudes towards multiple sexual partners (encouraged by easier and safer methods of birth-control) and more liberal views on homosexuality and illegal use of drugs have helped spread the often fatal AIDS, a threat currently reversing the trend to a more permissive society. Originally thought to be a problem only in the gay community, with evidence of no such limitation, sensible steps are being taken to encourage people to act in a way that might limit the spread of the latest in a long list of feared diseases. Sadly there is often a wide gap between encouragement and action, the width depending upon our attitudes to risk.

Genetic changes in a virus or bacteria – uncertain acts of nature – have rendered some established cures or preventive medicines ineffective. New strains of malaria are resistant to widely used anti-malaria drugs, so we are no longer certain of the general effectiveness of these drugs.

### New technology

Technological advances and social change, however obvious their benefits, bring new problems and sometimes new potential for catastrophe. The dam that controls flooding and brings hydroelectric power and water for domestic needs or irrigation – improving standards of living – may collapse and engulf those living in the valley below.

Jetliners carry the London businessman to a New York conference and return him, albeit somewhat jet-lagged, to his office within twenty-four hours. Flying has made a foreign holiday an annual event for millions who once would have seldom left their homeland. Yet we are shocked when hundreds die in a crash caused by mechanical or structural failure, by pilot error or by the mistake of an air-traffic controller.

On 31 August 1986, a light aircraft and a medium-sized jet collided over Los Angeles: more than 70 people died and wreckage destroyed houses. There have been aircrash death tolls seven times this number, but had the debris fallen on a city centre thoroughfare at a peak time the death role might have numbered thousands. The incident raised inevitable questions about air safety. All were thoroughly investigated, and the evidence carefully assessed by experts.

The public enquiry into the accident heard evidence of several contributory causes: unauthorized entry into controlled airspace by *more than one* light aircraft, possible deficiencies in the radar monitoring system, perhaps inadequate pilot lookout. Evidence was given that at twenty-three major U.S. airports a survey recorded 175 unauthorized intrusions by light aircraft in a six hour period! Alerted by the Los Angeles accident, and even before the public enquiry, the U.S. authorities had begun to tighten up on this kind of offender.

Air transport is a field where risk assessment based on hard evidence is well established, the hunt for evidence relentless. The psychological spin-off is increasing public confidence in air safety. Transport authorities generally have a good record in learning the lessons of catastrophe, and applying them to reduce the likelihood of repetition. Potentially hazardous industries have, with a few exceptions, been slower to learn from past mistakes.

### How logical?

How sensible are we about balancing risks and benefits? Our actions are sometimes influenced more by social factors like individual freedom or self-interest than by logical assessments. We know it is dangerous to drive under the influence of alcohol. How many of us can honestly say we have *never* driven with more than the legal limit of alcohol in our blood? Drunken driving (but not prosecutions for it) may be on the decrease, but it is not clear whether any decrease is because individuals are reacting logically to a proven danger. It is likely to be in part at any rate because the law now provides – and enforces – penalties that discourage the offence, and more effort is being made to detect and prosecute offenders.

### *Belt up!*

Some people are willing to put up with quantifiable, even considerable, risks to themselves if they feel that protective action denies a basic freedom.

In the U.K. each year some 6000 people (pedestrians, cyclists, drivers and passengers) are killed in road accidents; more than ten times that number are seriously injured. Because motor vehicles are important (the benefits are almost self-evident), nobody seriously suggests banning them. But would British Rail survive if they clocked up fatality rates near those attributable to motor vehicles?

Evidence that wearing car seat belts reduces the risk of death or serious injury has grown steadily. Many drivers, motivated by either prudence or fear, regarded a directive to 'belt up' a small price to pay in terms of liberty, but a vociferous minority, ruled by emotion, for long argued that, if they preferred not to wear a belt and were prepared to accept any increased risk (we are prone to believing accidents only happen to other people), that was their worry alone.

This is gross over-simplification. What about the cost to society of medical treatment for the injured, the pressure accident victims put on overloaded hospital facilities? It glosses over the social and economic implications for relatives (and indirectly to the community) of loss of a family breadwinner and the personal suffering of spouse and children of the deceased or injured.

## Emotional responses

Sadly, on important issues, many present-day politicians and activists, even some scientists, prefer emotive to reasoned argument, often backing their appeals with false claims. A growing army of professional lobbyists prey on this tendency to press their cases.

This may or may not be sinister; issues involving potential dangers are by nature emotive. Major controversies stir understandable emotions as diverse as fear, prejudice, sentimentality or patriotism; the result is often scant regard for rational interpretation of relevant information. Sincere and well-meaning evangelists and a few exhibitionists find they can better promote their cause, whether sound or merely a product of addled thinking, with a play on emotion rather than an appeal to reason. Irresponsible it may be, but it is also a very human attitude. Vested interests who plead their causes endlessly through professional lobbyists may have more sinister motives.

Relevant information needed for logical decisions is not always easy to come by and, with emotions highly charged, people are often reluctant to look for it; it may prove a pet hobby-horse unfit to ride. How easy also – deliberately or subconsciously – to distort or misinterpret evidence that contradicts cherished beliefs, or, more sinisterly and cynically, that might be economically damaging to vested interests. The tobacco industry took a lot of convincing that smoking is a health hazard.

We should not, indeed cannot, ignore emotions. But emotion and reason should not be locked in constant battle. Our most worthwhile emotional responses are those backed by reason, but more about psychological factors in the next chapter.

### *The rain it falleth . . .*

It is on environmental issues that emotions often run particularly high. The environment is a focus of rapidly growing concern, a trigger too for much irrational rhetoric. A plea for informed discussion came from Anthony Bradshaw, FRS:

> So many times, and I think it is more so recently, people discuss real problems of the environment in totally unreal ways, with emotion, without proper evidence, with speculation that has no foundation, and I think the important thing is . . . that we can hear all the evidence that there is, or almost all that there is.

Bradshaw's appeal was made at a symposium in Edinburgh in September 1985 (Anon., 1985). The subject, what is all too vaguely called *acid rain*. Bradshaw described the symposium as a unique opportunity to look at the subject in 'a properly meticulous manner', but more about this in Chapter 8.

### Benefit and risk

One very real problem in tackling uncertainty is how to quantify risks and benefits. Financial benefits may be easy to quantify; those involving quality of life, less so, since they may translate to different financial values for different individuals and for different communities. One person's gain may be another's loss. Risks are often hard to put into financial terms. If something is known about their pattern, they may be quantified in other ways, perhaps expressed as a death rate or as requiring a certain expenditure for a control measure of known efficiency.

Even with mundane problems like deciding optimum replacement policy for machines that have a variable lifespan, or the most efficient layout of a bank or air terminal, careful analysis is needed, and decisions are usually made to maximize profit, to minimize delays, or to avoid other causes of inconvenience or irritation, always taking hard financial realities into account in a way that often results in some compromise.

Quantitative analyses still have a key role in decision making for problems with a high social content, particularly environmental problems or those arising from natural or technological hazards. Such analyses cannot provide a complete answer, for the assessment of risks and benefits now contains an important socio-psychological component, a component that often dominates political discussion of hazards, perhaps reflecting a human trait. Many people find it easier to deploy and understand *qualitative* arguments than *quantitative* ones.

Even when we have evidence on which to base rational conclusions (effects of smoking on health is an example), heated, often irrational, argument still persists, argument not always (one is tempted to say seldom) divorced from vested interest.

If hard quantitative evidence is in short supply, we should search diligently for facts before making final and vital decisions; but tentative or provisional action may be needed while we collect such data. We are monitoring atmospheric pollution ever more carefully, aware of possible effects that may not be felt for many years. We have a lot to learn but we are taking controlling steps now, but often all too slowly.

## Our scope

To summarize, we shall be looking at

(i) psychological and emotive reactions to risks, in particular those associated with potential catastrophes;

(ii) misconceptions and prejudices which may arise from (a) lack of relevant data, (b) ignoring data or (c) misinterpreting data;

(iii) how to interpret factual evidence using concepts like *probability* and *utility* to assess potential benefits and risks;

(iv) problems in collecting, or selecting impartially, appropriate data to answer specific questions if these are not already available.

Sensible application of statistical methods needs more than an understanding of techniques. There are innumerable books that can help with that, some of which are listed in the Bibliography (p. 256). We also need a 'feeling' for data; that is something we try to develop in this book. That feeling helps us decide what data we need and how to use them for sensible decisions when there are uncertainties.

---

# OUR PERCEPTION OF RISK

### Types of uncertainty

There are two broad categories of uncertainty. The first, and simpler, can be described, and its problems dealt with, largely in what we call an *objective probability* framework. This is appropriate, for example, in an industrial context where we seek optimum procedures or want to assess some aspect of variability like the proportion of underweight items in a large mass-produced batch. Quality control methods enable us to monitor standards and ensure that market requirements are met. We may also want to study the likely impact of changes in systems, to see how we might improve efficiency or reduce costs; a technique called simulation that helps us do this is discussed in Chapter 10. In medicine a decision about the best drug to treat a disease may be based on statistical assessment of results of a well-planned clinical trial.

The second kind of uncertainty, broadly speaking, deals with potential hazards and exploring the unknown. There may still be a probabilistic element, but it is often subjective rather than objective; personal perceptions are an integral part of our attitude to risks and how we interpret them. In assessing benefits and risks, we mentioned utilities on p. 7. These vary from person to person. They are sometimes useful, although they may be ill-determined in situations where we have little reliable information.

Some problems of uncertainty do not fit neatly into one of the above classes, having elements from each. Improved technology and better quality control has improved [increased] air safety. We can predict with considerable accuracy 'safe' operating limits for engine components and airframes; here we deal with uncertainties of the first kind. However, acts of terrorism, human errors by engineers, pilots, or air-traffic controllers and mechanical failures of an unpredictable nature all introduce uncertainties of the second kind to air transport.

In this chapter we look at factors, many qualitative rather than quantitative, that influence our attitude towards risks and that are generally more relevant to the second kind of uncertainty.

## Our prejudices

Nearly 2000 pedestrians die on British roads each year; it is a bad year for commercial aviation when air fatalities reach that number worldwide. Why then are many people more fearful of flying than crossing the road or travelling by car? This is one of many situations where attitudes to risk do not always align with objective assessment. True, many of us cross the road several times a day and fly but once or twice a year, yet in any year a person selected at random is more likely to be killed crossing the road than to die in an air crash. Is it relevant that in the air we feel our fate is in the hands of other, rather remote, people (pilots, air-traffic controllers, maintenance engineers), whereas in crossing the road or driving ourselves, or being driven by a friend whose driving ability satisfies us, we have some control over our destiny?

*Objective* assessment of risk is based on known rates or a logically deduced probability. Popular perception is usually *subjective* and (often quite rightly) influenced by other factors, in particular any off-setting benefits. Purely objective assessments have been criticized on four grounds:

(i) they depend upon measurables (often just one), such as economic loss, morbidity or mortality, and ignore factors like inconvenience, irritation, fear, shock, or remote or long-term environmental degradation that are not easy to quantify;

(ii) assessments may be suspect because of lack of essential data or experience, making optimum decisions or reliable predictions virtually impossible;

(iii) assumptions in mathematical or statistical models may not truly reflect the real-world situation;

(iv) they have an abstract remoteness that often distances them both from those who create the risk (and prefer not to know) and authorities who should control the risk (but fear the political repercussions).

These criticisms are interrelated. The case histories in this book make it clear that shortage of relevant data or a lack of understanding of the scientific mechanisms that generate risk restricts the validity of many objective assessments (e.g. in atmospheric pollution or the spread mechanism of a disease).

Even if our store of *relevant* information is adequate it is unrealistic to ignore social attitudes that cause real concern about matters like the environment, the siting of hazardous plants or the dumping of noxious waste.

It is unfortunate that weaknesses in the objective approach have bred wider suspicion about how scientific assessments of risk are used. Are scientists employed by industry, government departments and other organizations producing unbiased reports? Do politicians interpret these for the common good or in the interests of favoured sectors? Even the most wily politician cannot please all the people all the time, but political lobbying and pressure politics are growth industries; there is a common belief that vested interests of the few often take priority over the good of the many in assessing dangers.

Jennifer Brown, a researcher on risk perception at Surrey University, elaborates on the theme of this section (Brown, 1987).

### The technology gap

Advanced technology has widened the attitude gap between those who reap the benefits of the new and those who fear, or perceive clearly, its risks. Incidents such as Bhopal, Chernobyl, the *Torrey Canyon*, Flixborough, Three Mile Island and the *Amoco Cadiz* have increased alarm about industrial accidents. These are names we associate with undesirable incidents, but how much do we know about what happened, or why, in each case? Our chief, and understandable, even if selfish, concern may be that the next accident could be at *our* neighbourhood petrochemical plant, at *our* nuclear power station or off *our* coast. How many lives will it claim?

Some factors influencing our reactions to risks are more rational than others. We have an inbuilt (some say primitive) fear of the unknown, a sometimes morbid enthusiasm for the dramatic, well exploited by the media. At the other extreme, contempt bred by familiarity may not be sensible, but it is a common reason for ignoring danger.

The concept of *utility* is relevant to the attitude of an individual, or group, to risks, but, even if we are given all relevant information, different individuals still arrive at different utilities for an analogous situation. With only partial information, there is likely to be an even greater variation between utilities assigned by individuals.

## Perceptions of danger

Dangers which, on an objective probability basis, are more likely than some unacceptable risks, are often tolerated because they have off-setting benefits. We accept relatively high road accident rates (compared to other forms of travel) because road travel is convenient and economic; most of us regard as reasonable the steps taken to reduce accidents to present levels. Introducing a universal speed limit of 15 mph (and enforcing it!) or building vast networks of overhead pedestrian bridges might lower pedestrian deaths, perhaps dramatically. We don't call for these things because the public at large consider the inconvenience and expense of a low speed limit and the difficulty of enforcing it, or the cost of a bridge network, would not be off-set by the (essentially unquantified) saving in life. The 'price' of the bridges may be regarded as too high in more than straight economic terms; it may include the environmental consideration that most overhead bridges are eyesores.

## Scales of measurement

Off-setting benefits is only one reason why it is not appropriate to compare risks on a single scale of 'probabilities'. Single scales are also unsatisfactory because relative risks depend critically on the frame of reference; for several different hazards, numbers of deaths per annum will in general give a different ordering from deaths per unit distance travelled or number of hours' exposure to risk. For example, if we calculate death per unit time spent on a given activity, being burnt alive in the home is less likely than being killed when rock climbing. But, if we count simply deaths per year (irrespective of how much time, if any, each person spends on the relevant 'activity'), fire in the home is a more likely cause of death.

Table 2.1 demonstrates some of the difficulties in comparing risks in different fields. The table is confined to one risk, *death*; it is subdivided into four basically different kinds of hazard, labelled (a)–(d).

## A post-mortem on the table

An obvious problem for comparisons between different subdivisions (a) to (d) in Table 2.1 is that the bases for the rates in each are different. Only for categories (a) and (b) are they rather similar (effectively per million at risk), but even these are not completely identical. Industrial risk is more or less uniformly spread over a period, but medical risk is a one-off affair for most people, repetitive for an unfortunate few; for anaesthesia, for example, a person who undergoes one operation in the observation interval is recorded once in the data, but a number of people, particularly the very sick, may have several operations and therefore are exposed several times to this risk. A person who has three operations in a year may have a serious health problem and so be at especially high risk under an anaesthetic.

Category (c) is based on distance and category (d) on time of exposure, both quite logical for the risks in question (although not the only sensible possibilities). Using a common base for all

*Table 2.1.* Death rates for various risks

| | |
|---|---:|
| (a) Medical Risks. Deaths per million cases | |
| Vaccination (England & Wales, 1967–76) | 1 |
| Surgical anaesthesia (England & Wales, 1970–73) | 40 |
| Childbearing (England & Wales, 1974–76) | 100 |
| Needle biopsy of liver (period not stated) | 200 |
| (b) Industrial accidents, U.K. (1974–78). Deaths per million at risk. | |
| Clothing and footwear manufacture | 5 |
| Vehicle manufacture | 15 |
| Chemical and allied industries | 85 |
| Railway staff | 180 |
| Coal mines | 210 |
| (c) Travel. Deaths per billion kilometres travelled (1972–76). | |
| Railway passengers | 0.45 |
| Scheduled airline passengers | 1.40 |
| Car or taxi | 8 |
| Pedal cyclists | 85 |
| (d) Sport. Deaths per million participant hours, U.K. | |
| Amateur boxing (1946–62) | 0.5 |
| Canoeing (1960–62) | 10 |
| Rock climbing (1961) | 40 |
| Scuba diving (1970–79) | 220 |
| Hang gliding (1977–79) | 1500 |

*Source: Risk Assessment.* A Study Group Report. The Royal Society, 1983.

four subsections would be theoretically possible, but not very sensible. There are objections to expressing sporting accidents simply per million participants; clearly, time of exposure to risk is pertinent here, as is acknowledged in section (d). Even in one subsection of the table, one might diverge by using different bases for each risk; for example, in (c) we could have based accident data on numbers of cyclists, car owners, airline passengers, etc., and expressed rates per million of each involved.

All this means it is virtually impossible to compare a medical risk with that of rock climbing by using a numerical basis that provides just a single figure for each.

## Comparable rates with different meanings

Table 2.1 shows death rates for needle biopsy of the liver and for coal mine accidents are almost the same, but the implications are different. A needle biopsy is performed for diagnostic or treatment purposes only when a doctor considers it necessary. If biopsies were abandoned, there would probably be an increased death rate among patients who, under accepted practice, are now given biopsies. Instead of 200 per million, the rate might increase to 500 or 1000 – even to 10 000 – per million for these people! Clearly, the biopsies have a high benefit component, and doctors are trying to reduce the death rate from them. Such deaths are often the result of haemorrhage, so some who die have a liver malfunction together with a blood coagulation problem. Death rate for biopsies is probably dropping as special precautions are taken with high-risk cases.

Coal mining accidents might be reduced by more stringent safety precautions, though the industry is safety conscious and on a cost–benefit basis the rate is generally regarded as *acceptable*, in the sense that additional expenditure on safety is unlikely to reduce the rate appreciably and might make many mining operations uneconomic. There is also a voluntary element: an individual who regards the rate as unacceptable would presumably leave (or not enter) the industry.

## No equivalence

There is no way we can sensibly compare a rate of 1500 deaths per million *participant hours* of hang gliding with forty deaths per million *cases* under anaesthesia. These crude rates may also hide large differences for particular subgroups. Deaths under anaesthetics are more common among elderly patients with cardiac irregularities undergoing major surgery than they are for minor routine operations on basically healthy younger patients. Hang gliding accidents among members of well-organized clubs providing proper instruction are less common than for poorly organized groups. Indeed, alarm at the high accident rate in the early days of this sport (covered by the Table 2.1 data) led to a tightening in official controls, and to the hang gliding movement itself introducing better codes of practice.

## Paucity of data and subjectivity

Even where comparisons of relative risks make sense (e.g. between two industries or between two sports), popular perceptions of risk do not always accord with relevant data.

Studies, especially in the U.S.A., have shown a tendency to overestimate rare adversities and to underestimate more common ones. In one U.S. study forty hazards were compared after people had been told the annual numbers of road deaths. The tendency was to overestimate deaths due to pregnancy, childbirth, abortion and cancer (relatively lower), and to underestimate deaths due to smallpox (still fairly common in parts of the world at the time of the study) or being struck by lightning. This tendency to 'pull' estimates towards a known basic figure such as numbers of road deaths is called an anchorage effect, and we say more about it in the next chapter.

Compared with known hazards, death rates reported in U.S. newspapers bear little resemblance to actual rates. Murders, natural disasters and a variety of accidents are over-reported relative to hazard-induced disease (individual or epidemic), unless these are something new like AIDS, or the victims are well known.

This media attitude to hazards may influence public perception. Murders are often thought to be as common as cerebral haemorrhages (strokes); yet, even in countries with a high murder rate, strokes are usually up to ten times more common. People dying from a stroke will only rate a mention in their local weekly if they are prominent citizens and in the national press if they be Royalty, a cabinet minister or a pop star; even then, a stroke as cause of death may not be specifically mentioned. Is this a chicken and egg situation? Does media coverage mould public perception of relative risks or does media coverage reflect that perception?

Real as a media effect may be, it is only one factor influencing subjective attitudes. We may ask: Is the risk voluntary or involuntary? A threat to society at large or only to a particular subgroup (e.g. those working in a particular industry)?

## Acceptable risk

Public policy towards risk usually acknowledges, tacitly if not overtly, that complete elimination of most risks is impossible (sometimes undesirable). Yet complete elimination is often the aim of (usually ill-informed) pressure groups.

There is an important difference between *accepted* and *acceptable* risk. Sulphur emissions from power stations is an *accepted* risk, but many people (probably a majority) regard present levels in many countries (the U.K. included) as *unacceptable*. The consensus of opinion is currently moving to a belief that we should set maximum risks at *acceptable* rather than *accepted* levels.

Legislation to reduce risk is therefore increasingly being based on some *maximum acceptable* level. In almost all cases this depends on the corresponding benefit. A real difficulty is to get consensus on what is a reasonable *maximum acceptable* level.

The Royal Society report on risk assessment (Anon., 1983) suggests that a shift in majority attitudes towards responsibility may be summed up this way:

> They (the majority of people) have generally assessed the world as a fairly hazardous place and have looked mainly to religion to provide meaning and explanation for this characteristic and to find particular reasons for the disasters that befall them individually and collectively. It is possible that some part of this omniscience has been transferred, in the public's mind, to Government . . . Governments are now seen to have a plain duty . . . to apply themselves explicitly to making the environment safe.

## *The nuclear debate*

Nuclear power is a matter where there is a wide diversity of opinion about what is acceptable. Many of its opponents regard the only acceptable level of risk as zero – implying no nuclear power stations. Proponents of nuclear power consider that present safeguards (at least those in the U.K.) imply less risk from nuclear power than from other forms of energy and therefore that what risk there is should be accepted.

The nuclear power controversy has intrigued psychologists studying personal attitudes to risk. They have suggested that one reason the nuclear debate generates more heat than light is because the two sides have little common ground in their assessments of risks and benefits. Each regards different factors as important and virtually discounts the things the other side regards as vital. Nuclear experts and those with a commercial interest in energy stress the low costs and the currently low accident rate. They discount the possibility of another disaster of Chernobyl or greater dimension as infinitesimally small. The anti-nuclear lobby dismiss the energy arguments as irrelevant, claiming there are ample preferable alternative sources of power and asserting that, however small the probability of a major disaster, its consequence would be so grave it cannot be ignored. They also advance dangers of terrorist action and question marks over disposal of nuclear waste as factors ignored or down-weighted by the pro-nuclear lobby.

Psychologists have suggested that, while nuclear scientists and informed but dedicated environmentalists have respect for the views of each other (even if they do not agree with them), there is suspicion on both sides about what each regards as the 'fringe element' of the opposition; these include on one side pro-nuclear politicians, economists and energy experts who are not specialists on specifically nuclear matters, and on the other politically motivated or fanatical environmentalists. In the U.K., at least, such attitudes may be a legacy of the unsympathetic attitude (now being slowly modified) of the electricity industry and Government to issues like acid rain.

### Risks and probability

Yet another impediment exists to arranging risks on a simple scale of increasing probability: one hazard may lead to several different risks, risks that cannot be ordered unequivocally in order of importance. While death is generally rated more serious than non-terminal illness, some fates (e.g. irreversible brain damage or total paralysis from the neck downward) are widely regarded as 'worse than death'. Would thalidomide have attracted the attention it did if it had only led to a few deaths due to cardiac failure among those who took it instead of inflicting lifelong physical malformation on unborn babies (innocent victims)? Nobody knows how often euthanasia is practised, but that it is, and has considerable backing, suggests a substantial body of opinion subscribing to the *fate worse than death* syndrome.

### Alternative risks

We often prefer a high probability of minor danger or inconvenience to a small probability of serious risk.

If a chemical process produces fumes that may prove irritating (but not fatal or seriously debilitating) to 50% of workers at a plant or to nearby residents, most people would agree this is in no sense a worse risk than that from introducing an alternative process that avoided these fumes, but carried a danger that one worker in 500 might develop cancer of the pancreas as a result of exposure. It is a separate question whether even the first process should be tolerated; a decision would in practice be made by considering the amount of irritation caused and the economic benefits (jobs, etc.) provided by the plant.

### One man's meat . . .

There are often relevant, though essentially different, populations for whom the risk–benefit equations are not the same, and this affects attitudes. A chemical plant producing unpleasant fumes may pay its workers well. If the fumes are not a serious health hazard, the workers in the plant might gladly put up with them to earn good wages. The plant has a high utility for these workers; yet it may be unacceptable to environmentalists or farmers who see the fumes damaging vegetation over a wide area. There is no magic formula to achieve equity in this case. As in the nuclear energy controversy, one party tends to dismiss the damage as irrelevant; the other discounts the benefit as negligible (or irrelevant to them!). A psychologist might ascribe both attitudes to selfishness.

### Building up our perceptions

With many different types of risk, it is not surprising our perceptions are governed by combining, in a complex and individual way, information from direct experience, published data, reports received second-hand (e.g. via the media) plus our own 'scaling' of accident potential relative to our personal experience and instincts. If we have a fear of drowning, but no fear of heights, the risk of going to sea in a small boat seems greater than that of rock climbing.

Psychologists have tried to assess the relative importance of various influences of this type. They have looked at how people view different risks: for example, whether they see them as personal or as threats to society; whether they are voluntary, reasonably avoidable, and so on; and also how like or different various risks are.

Perceptions change with time, partly because of additional information, but also because risks themselves change. Nuclear power and liquefied natural gas currently give more concern than air travel. Thirty years ago, the first two were seldom heard of; air travel was then, as objective statistics show, less safe than it is now.

### Comparing risks

A seminar at Sunningdale in 1979 sponsored by the Department of the Environment listed competing criteria by which people assess risk. Their conclusions were as follows:

1. Concentrated risks (Bhopal, Jumbo jet crash) are regarded as worse than diffuse risks (road accidents, climbing accidents).

2. Risks to non-beneficiaries (people living *near* a nuclear power station or in a valley below a dam, or the crew of a lifeboat) are regarded as worse than risks to beneficiaries (workers in a power station whether nuclear or hydroelectric at the foot of a dam, or fishermen).

3. Involuntary risks (poisoning from contaminated food) are regarded as worse than voluntary risks (over-indulgence in alcohol).

4. Imposed risks (reducing the staff of the fire service to save money) are less acceptable than risks undertaken for self-protection (use of a well-tested vaccine, even if it is known there is a small probability of a serious side effect).

5. Isolated risks with no compensating benefit (falling through a broken manhole cover in the street) are less acceptable than risks incidental to an otherwise largely beneficial context (a house being burnt down by a fire in a deep-fry cooker).

6. Immediate hazards (electrical faults in a new washing machine) are seen as worse than deferred hazards (faults resulting from poor maintenance).

7. Risks from unfamiliar or unnatural hazards (a new food additive) are regarded as worse than those from familiar or natural hazards (crossing a busy road).

8. Risks resulting from secret activities (chemical leaks from a defence establishment) are regarded as worse than those associated with open activities (mining or smelting).

9. Risks evaluated by groups suspected of partiality (e.g. based on an industry's own assessment of its safety) are regarded as worse than similar risks evaluated by impartial groups (e.g. a Royal Commission or a committee of enquiry set up by a learned society).

10. Risks that someone else has to pay to put right are regarded as worse than risks for which people have themselves to pay to remedy.

This last contrast may seem surprising, but we are often reluctant to pay for something we feel is of little benefit to us. If an outside staircase is dangerous, we may not bother to provide a handrail.

We know the danger and take care; it is just too bad for those unfamiliar with the hazard. The story might be different if our landlord had to pay to put the matter right, or if there were no handrail on a public stairway.

The Royal Society report suggested an interesting addition to this list, the *named life* factor, pointing out that more effort is sometimes put into an air/sea rescue, and more concern felt, if an important person is involved than if the life of an unknown is at risk.

## Research into risk perception

There is a lot of interest in why, and how, attitudes to hazards differ. We review briefly some current thinking about this.

*Risks to society.* In risks affecting society as a whole, personal factors play a minor role in judgement, because the impact of decisions is more remote; however, the way information is put to us may affect our attitude. The increased risk of any one person dying of skin cancer if we reduce the stratospheric ozone layer will be small. If told it is likely to be 1 in 5000 in our region, an individual may be prepared to ignore it; but if we are told the number of deaths worldwide is expected to increase by 10 000 per year, we are more likely to feel something should be done about it. We take more notice of absolute numbers when proportional changes are small, so stating a risk in absolute terms brings greater public awareness than saying there will be an increase of, say, 0.02%, or even 1 or 2%.

*Immediate risks.* People are generally more averse to risks that are immediate than to those long delayed. Examples of the former are skiing, air travel and crossing the road, and of the latter, smoking, asbestos dust and food preservatives. However, recent research indicates that a delayed effect in itself has less influence on our perception than was once thought. Any differences may be due more to lack of familiarity with a hazard, this often, but not always, being associated with a delayed effect.

*Catastrophes.* Natural disasters form a special category. People are usually fatalistic about these, and better early warning systems, in the hope that people will act on them, may be the only useful preventive step. People usually resent compulsory removal from the shadow of a volcano or a flood or earthquake zone. Many of us who are not ourselves farmers living in an area subject to flood or drought may wonder why a farmer facing this problem does not move. He may decide the benefits of owning a home he likes, being near friends and doubts about being accepted in another community outweigh the adverse effects of flood or drought. He probably subconsciously justifies his decision by down-weighting the degree of risks or by rating highly his utility for living where he does.

### The numbers game

Green and Brown (1978) have studied people's ordering of various risks on the bases of *personal safety* and *threat to society*. In both categories they found people were influenced in their assessment by the *kill size* or number of deaths or injuries that might result from a single incident.

Loss of a whole crew of a lifeboat has more impact than the same number of fishermen lost in a number of incidents spread over months.

Several workers have suggested that we probably react to a one-off mass loss of life as though the square, or even the cube, of that number of lives had been lost in single incidents (e.g. if a lifeboat crew of six is killed, we react as we might if thirty-six or more fishermen were lost in a number of isolated incidents). There is probably another factor at work in this example, namely, that we react more emotionally to accidents to those we feel are doing a worthy job. In a 1987 alert in the Scottish mountains, a member of a rescue team was killed while trying to recover the body of a dead climber. There was wide media coverage and public sympathy for the member of the rescue team; little for the climber. This is perhaps a 'good citizenship' effect something like the 'named life' factor. Both attitudes are in conflict with the Christian view (and that of many religions) that each life is of equal importance. It reflects our tendency to admire heroism while being critical of foolhardiness and lack of foresight that place others (e.g. members of rescue services) in danger.

## Distance lends enchantment

Attitudes to tragedy are influenced by geographical or social distance. A flood in Pakistan may well bring less reaction in the U.K. than when a cross-channel ferry capsizes with only one-twentieth the loss of life. Perhaps in this example there is an additional factor that we may be more ready to accept natural catastrophe than that due to human error or technological failure, but that does not explain why the sinking of the Philippine ferry *Dona Paz* with a loss of over 1500 lives in horrifying circumstances on 21 December 1987 was not even the lead news item in *The Times* the following morning, and that a day when competition from other items was far from strong.

Studies have shown that, while an expert's perception of risks is often closely correlated with the numbers of deaths relative to those from other risks in some sense related, this is not generally true for the public at large, probably because of less familiarity with the true figures, the influence of media reporting, lobbying by interested parties and so on.

## *. . . or disenchantment?*

An interesting study at a Continental nuclear plant showed that anxiety was not a simple function of distance from the plant. Anxiety was greatest among those living between 2 and 4 km from the plant. This could reflect a tendency for those with real anxiety to refuse to live too close. It may also mean that those living near the plant (workers, etc.) play down dangers because of the lure of high pay (or is it a case of familiarity breeding contempt?).

## *Cost effectiveness*

Publicity given to dramatic accidents may result in costly, but not always cost-effective, steps to prevent a repetition. This may reflect over-reaction.

Tragedies such as the thalidomide affair and other incidents involving unfortunate side effects of drugs led to a large invest-ment in drug screening, the pharmaceutical industry and govern-ments responding to a public demand for more stringent testing. Many more lives (but those of different people) might be saved if half the money now devoted to drug research and testing were channelled (effectively) into famine relief.

Mandatory fire precautions in the U.K. cost far more per life saved than would the corresponding amount diverted to improv-ing certain hospital facilities.

It has been claimed that rehousing costs and 'essential' modifica-tions following the Ronan Point disaster in London (in which an explosion in a high-rise building exposed a serious design weak-ness) probably represented £20 million per life saved. Any such figure itself carries a high uncertainty element. It is relatively easy to identify housing blocks liable to structural failures of the type involved at Ronan Point, but virtually impossible to say in how many cases a corresponding (or more disastrous failure) would ever take place. It is equally difficult to say whether less expensive corrective measures might not have been just as effective.

Even if expenditure in the Ronan Point aftermath was not cost effective nationally or globally, it was important for residents of similar blocks. They were entitled to alternative housing or action to make their homes safe. Their utility rating for these safety measures was high.

When popular perception and objective assessment differ, the fault is not always on the side of an inexpert public: the objective assessment may not be as good as it ought to be. In Chapter 8 we examine disagreements about causes and effects of atmospheric pollution. The U.K. Central Electricity Generating Board (CEGB) experts were long the 'odd men out'; only recently have they accepted unreservedly that CEGB stations are major contributors to damaging acid rain in Scandinavia. Here, the experts have been drawn into line with lay thinking.

## Was Franklin right?

Uncertainty and risk are with us all the time, even if not quite as all-pervasive as Benjamin Franklin claimed on p. 3. It is better to face the realities than to sweep uncertainty under the carpet. It was perhaps the greatest weakness of classical science that it devoted much of its effort to ignoring uncertainty. This is still a flaw in some of today's scientific and technological work.

## Summary

1. There are two basic types of uncertainty: (i) probabilistic and (ii) largely unpredicted or unpredictable hazards. Some risks are a mixture of the two.

2. Human perception of risk may not correspond to objective assessment, and a single scale for ordering risks is not always appropriate; also ordering may differ with different bases (e.g. deaths per 1000 hours exposure may not give the same ordering as deaths per annum).

3. Different utilities for individuals or groups result in different perceptions of, and responses to, potential hazards.

4. We should reduce risks to acceptable rather than stay at accepted levels.

5. Individual perception of risk depends on competing criteria: for example, numbers at risk, compensating benefits, whether voluntary or involuntary, immediate or delayed, familiarity, who pays and so on.

6. Measures to reduce risk may not always be highly cost effective, but they may be dictated by social demands.

## CHAPTER 3

## ABERRATIONS AND DISTORTIONS

### Numbers and intuition

In Chapter 2 we looked at uncertainty and risk in a qualitative way. Studying quantitative aspects is more exacting. One source of difficulty, little discussed in most formal treatments, is the danger of wrong assessment of, or erroneous deductions from, data. In this chapter we look at some of the pitfalls in how we perceive and manipulate data.

Good decisions are based on good information, but we must use it carefully. In particular, treat intuition with caution; it is a good servant, a bad master. It is easy to back wrong hunches or use flawed logic. With little data (however good) or only a brief description of a situation, we tend to make quick assessments and draw what seem obvious conclusions. Here are a few tricks intuition may play.

### Arithmetical intuition: how good?

*Example 3.1.* Look at the following calculation for five seconds – no more – and write down your estimate of the answer. Calculate

$$10 \times 9 \times 8 \times 7 \times 6 \times 5 \times 4 \times 3 \times 2 \times 1.$$

Now turn to p. 52 to see the correct answer. Are you surprised? Did you expect it to be so large?

How did you get your estimate? Guess? Or in five seconds do the first few multiplications, i.e. $10 \times 9 = 90$, $90 \times 8 = 720$, $720 \times 7$ is approximately 5000, then estimate the final answer after noting that we multiply by smaller numbers each time, thinking it would be about 50 000, perhaps 100 000? That is what most of us do.

Had we posed the problem as: calculate

$$1 \times 2 \times 3 \times 4 \times 5 \times 6 \times 7 \times 8 \times 9 \times 10,$$

the chances are your guess would have been lower, since multiplying the first few numbers gives $1 \times 2 = 2$, $2 \times 3 = 6$, $6 \times 4 = 24$, $24 \times 5 = 120$. The products are getting bigger, but most of us tend to underestimate how fast they grow: guesses between 1000 and 20 000 are common.

### An experimental result

Tversky and Kahneman (1974) divided a class of high-school pupils into two groups. The first were asked to estimate (or guess), within five seconds, the value of the product $8 \times 7 \times 6 \times 5 \times 4 \times 3 \times 2 \times 1$; the second, that of the same product in reverse order, i.e. $1 \times 2 \times 3 \times 4 \times 5 \times 6 \times 7 \times 8$. The median estimate (i.e. the value such that half the group gave that or a lower value and half that or a higher value) was 2250 in the first case and 512 in the second. The correct answer is 40 320. The tendency to underestimate the product of all integers between 1 and 8 or between 1 and 10 (no matter in what order they are given) illustrates a human trait. Given only an initial estimate based on a calculation using part of the data, we tend not to adjust it sufficiently.

### Anchorage effects

Getting a lower estimate for $1 \times 2 \times 3 \times \cdots \times 8$ than we do for $8 \times 7 \times 6 \times \cdots \times 1$ is a secondary but related characteristic called an *anchorage effect*; these are common in mental assessments. In this example, if the numbers are given in ascending order, $1 \times 2 \times 3 \times \cdots$, we start with a small product estimate and *anchor* on that. Given the descending order, $8 \times 7 \times 6 \times \cdots$, we *anchor* on the larger first estimate.

Tversky and Kahneman also asked students in two groups to estimate the number of African states in the U.N. One group were asked to say whether they thought it more or less than sixty-five, then estimate the number. Another group were asked the same question with ten in place of sixty-five. The median estimate for the first group was forty-five and for the second twenty-five. This was clearly an anchorage effect.

### Anchorage and risks

American researchers led by Sara Lichtenstein (Lichtenstein *et al.*, 1978) found an anchorage effect in people's concept of risk. Two groups were asked to estimate the frequencies of deaths in the U.S.A. from each of forty causes. One group was given the prior information that about 50 000 people die each year in road accidents; the other that about 1000 die from electrocution. These are correct figures. A larger anchor number increased respondents' estimates of frequencies for other causes; the smallest estimates given by members of the first group were mostly about five times as large as the smallest estimates given by members of the second group.

### A further test

Here are two more intuition tests. Take as long as you like, but decide your answers before checking on p. 52.

*Example 3.2.*    In English does the letter r occur more commonly at the beginning of words of three or more letters, or in the third position?

*Example 3.3.*    What is the smallest number of people (none of whom are twins) at a party if there is to be a better than even chance that at least two will have a birthday anniversary on the same date (i.e. are born on the same day of the year but not necessarily in the same year). Pick your answer from one of the following: 10, 17, 23, 29, 45, 52, 57, 63, 81, 121, 129, 141, 152, 183, 212, 334, 421.

### Influence of background information

Differing amounts of background information may influence responses. Baruch Fischoff and a colleague asked a number of people to estimate the annual death rate from influenza per 100 000 who catch the disease: the mean response was 393 deaths per 100 000 cases. Another group were told that 80 million people catch influenza in a normal year and asked to estimate the number that die: the mean response was 4800. Now 4800 per 80 million is equivalent to six per 100 000; less than one-sixtieth of the mean estimate, 393 per 100 000, given by the first group who did not know the total number of cases. This and other examples are quoted by Fischoff *et al.* (1982).

### *Neglecting information*

*Example 3.4.*    We are told that in a community 80% of the people are white and 20% black. Clearly, if one person is selected purely by chance (*at random* (p. 107) is the technical term), the odds are 4 to 1 in favour of that person being white (i.e. the probability of being white is $\frac{4}{5}$). There is no catch here.

Many people ignore these odds if they are given additional, yet virtually useless, information about the person selected. For example, if told the person were male, aged twenty-seven, about average height, wore a pullover and grey slacks, spoke with a slight lisp and was accompanied by a small child, people often assert the man is *equally likely* to be black or white, an answer that reflects lack of useful information about race in the personal details. Why not ignore this useless material and revert to the prior information that only 20% of the population are black?

We often tend intuitively to give more weight to information, however useless, about one selected person, than we do to *relevant* information about a complete population.

### *A bridge problem*

We sometimes overlook information because we do not see its relevance.

*Example 3.5.*    There are two bridges across a river. About twenty cars cross bridge A each day; about 120 cars cross the busier bridge B. Roughly the same number cross north to south and south to north each day on each bridge, but there is some day-to-day fluctuation. For each bridge records are kept during one year of the numbers of days on which more than 60% of the cars cross from south to north. Which bridge would you expect to show the greatest number of days exhibiting this phenomenon? Assume the bridges are reasonably far apart and on any day trends in traffic flow on one are independent of those on the other. Possible answers are: (i) bridge A; (ii) bridge B; (iii) roughly the same for each bridge. Decide, then look at the answer on p. 52.

## Further examples

*Example 3.6.*    A coin is tossed seven times. The outcomes, heads (H) or tails (T), are recorded in order for the tosses. Which of the following sequences do you consider the most likely?

H T T H T H H
H H H T T T H
T T T T T T T

Decide, then look at the answer on p. 53.

*Example 3.7.*    Committees of size 3, 4, 5, 6, 7, 8 or 9 are to be selected from a group of twelve people. How many different committees can be formed depends on the size of committee (committees are different if their composition differs by at least one person). For what size can we get the greatest number of different committees? For what size the least number?

What do you think? 'Most of size 3; fewest of size 9' is a popular answer. It is easy to visualize a lot of different committees of size 3; harder to picture different committees of size 9. In Chapter 5 we find that there are 220 different committees of size 3 and the same number of size 9. Size 6 gives the most – 924 different committees, over four times the number of size 3 or 9. Even if we do not know how to calculate the numbers of committees of size 3 and 9, it is easy to see the numbers must be the same; this is because each different selection of 3 leaves aside a different group of 9. Each of these forms a 'committee' of size 9, each 'pairing' with one of the different committees of 3. Mathematicians speak of a *one-to-one correspondence* between committees of the two sizes.

## Communicating

Impact of information depends on how it is presented. If a spokesman says the Government will spend £100 million on new jet fighter planes, this will probably mean little to most people: governments always talk in millions and aircraft are expensive. But tell them this amount could provide every man, woman and child in the U.K. with a pint of milk a day for a week, or that it would chop about £6 off the television licence fee if it were given to the BBC, or that it is less than £2 per head of population, and then people have a perspective (something with which they are familiar) for interpreting the amount involved.

## Visual distortion

In large semi-arid tracts of the Australian Northern Territory, the two most obvious forms of land-based animal life are kangaroos (large marsupials) and termites (small insects). Suppose we were able to collect together and weigh both all the termites and all the kangaroos on a large area. How do you think the total weights would compare? Would the total weight of all the termites be about a thousandth of that of the kangaroos, or a hundredth, or a tenth, or might there be about equal weights of each?

To find out, a group of scientists got accurate estimates of the weights of each per unit area of land. For kangaroos this was about 1 kg per hectare; for termites about 700 kg per hectare! A result that surprised even those who knew the country well. Our visual concept distorts the true picture. Kangaroos are large and obvious: we cannot miss seeing them, even a long way off. Termites, on the other hand, are small and somewhat inconsequential (until they invade one's larder): we only see those that are close to us. Also, because they live in mounds, for every one we see there are probably a million or more we don't.

## Pitfalls in calculations

Our examples so far have been of situations where intuition may fail because our 'hunches' or our perceptions are wrong. Errors also occur when we think we are carrying out the right calculation, but do not fully appreciate the logic of a problem.

### A classic averaging trap

To work out the average speed for a motor journey the *correct* rule is

*divide total distance travelled by total time taken.*

*Example 3.8.* I drive from my home to Glasgow, a distance of 90 miles, at a speed of 45 mph, and return at a speed of 30 mph. Can I short-cut my calculations and assert the average speed for the whole journey is 37.5 mph (i.e. the average of the two rates 30 and 45 mph)? Let us check. The trip to Glasgow takes 2 hours (the time to travel 90 miles at 45 mph); the return journey takes 3 hours (the time to travel 90 miles at 30 mph). So the total journey time is $2 + 3 = 5$ hours for distance of $90 + 90 = 180$ miles. My average speed is thus $180/5 = 36$ mph, *not* 37.5 mph.

What has gone wrong? That value 37.5 is an *arithmetic mean* of the each-way speeds 30 and 45 mph. Although it is the most commonly used average, the arithmetic mean is not appropriate here. This example, and our next, show that, though arithmetic means are important, there are pitfalls with their use.

### Hardly cricket

This example is about cricket. If you are not an addict, all you need know is that a *bowling average* is calculated by dividing the total runs scored by batsmen against a bowler by the number of

*Table* 3.1. Bowling analysis, fifteen games

|  | Bloggs | Smythe |
| --- | --- | --- |
| Runs scored: | 210 | 214 |
| No. dismissed: | 30 | 30 |
| Average: | 7.00 | 7.13 |

batsmen that the bowler dismisses (also referred to as the *number of wickets taken*). A bowler likes a low average, for his aim is to dismiss batsmen for as few runs as possible. The average here is an *arithmetic mean*, but beware of some of its properties.

*Example 3.9.* A cricket club gives a cup to the bowler with the best (i.e. lowest) average. Last season there was a neck and neck contest between Bloggs and Smythe. The secretary used information from the scorebook for all fifteen matches on his printed fixture list to compile Table 3.1.

The averages (arithmetic means) are obtained by dividing the runs scored by the number dismissed (e.g. for Bloggs it is 210/30 = 7.00). The secretary concluded Bloggs had won the cup by a whisker. Then he suddenly remembered that he had not included figures for an extra match that was not on the fixture list. Figures for that match are given in Table 3.2.

*Table* 3.2. Bowling analysis, extra game

|  | Bloggs | Smythe |
| --- | --- | --- |
| Runs scored: | 70 | 39 |
| No. dismissed: | 5 | 2 |
| Average: | 14.00 | 19.50 |

Clearly Bloggs had the better average in this game also, so deserved the cup. Just for the record the secretary added runs scored and numbers dismissed in this game to those in Table 3.1 and calculated the final averages for all sixteen games, producing Table 3.3, where 280 = 210 + 70, 253 = 214 + 39, and so on.

*Table 3.3.* Bowling analysis, all games

|              | Bloggs | Smythe |
|--------------|--------|--------|
| Runs scored: | 280    | 253    |
| No. dismissed: | 35   | 32     |
| Average:     | 8.00   | 7.90   |

What's this? Smythe has the better average! Is something wrong with the secretary's pocket calculator? Check for yourself. Do you agree with Table 3.3?

Have we all got faulty calculators? No, this is just another deception with averages. Compared with the fifteen matches providing the data for Table 3.1, in the one remaining match (Table 3.2) both bowlers have performed relatively poorly in the sense that each has a higher average. In this one match Smythe had done even worse than Bloggs in the senses that (i) he has taken fewer wickets and (ii) has a higher average (Table 3.2 again). Smythe has done better in one way: fewer runs have been scored off him than off Bloggs – 39 compared with 70. Combining the results with those from the other games and calculating averages anew, the arithmetic mean penalizes Bloggs more heavily for those extra runs, despite his having taken more wickets.

Look at it this way. In the extra match Bloggs has *five* relatively poor dismissing performances while Smythe has only *two*. In the final analysis these two get less weight because they are two out of thirty-two dismissals, or only *one-sixteenth* of all Smythe's dismissals. Bloggs takes five out of his thirty-five wickets, that is *one-seventh* of all his wickets in this match. The arithmetic mean takes account of these proportions, using them as what are called *weights* in calculating the overall average.

## Testing a new drug

Paradoxes like the one with bowling averages arise in more important contexts. Direct additions of runs and numbers of dismissals was appropriate in Example 3.9, but it is sometimes misleading to combine data by direct addition: we must look at the separate parts, or combine them in some other way.

*Example 3.10.* The standard treatment for a respiratory disease cures less than half of all patients. Its efficacy depends on the severity of the complaint and on whether the patient lives in the city (where atmospheric pollution aggravates the condition) or in the country (where treatment is more likely to be effective because of cleaner air). Will a new drug do better?

To make a fair comparison health officers test the new drug both in city and country. They decide the trial should cover 2000 patients in the city and 1000 in the country; records show that these are about the numbers of cases that can be expected in three months. They would like to give half the patients in each group the old treatment and half the new, but the new treatment is harder to administer, requiring several visits from the doctor. Country visits are time-consuming and, since doctors are busy people, the organizers decide to try the new treatment on 70% of city and 30% of country cases, all other patients to receive the more easily administered standard treatment. A statistician confirms this is not unreasonable. He also advises a *random* choice of patients who are to receive each drug. This avoids a possible bias. Without this precaution some doctors might, for example, give the new drug only to the worst cases because they know the old one is practically useless for such sufferers and hope the new one might do better. If patients were selected this way, it would be impossible to interpret sensibly what might happen in more general use. We elaborate on the term *random* on p. 107. Essentially it means each patient has the same opportunity as any other of being chosen for treatment with the new drug; those not so selected are given the standard one. The results are summarized in Table 3.4.

Trying to be helpful, the health officer combined city and country figures and just gave the statistician Table 3.5.

The latter easily calculated that, for the standard drug, 450 cures in a set of 1300 patients represented a cure rate of $\frac{450}{1300} \times 100 = 34.61\%$; for the new drug the rate was only

*Table* 3.4. Numbers responding to two drugs

|  | City | | Country | |
|---|---|---|---|---|
|  | Standard | New | Standard | New |
| No effect: | 500 | 1050 | 350 | 120 |
| Cure: | 100 | 350 | 350 | 180 |
| Total: | 600 | 1400 | 700 | 300 |

*Table* 3.5. Total numbers responding to two drugs

|  | Standard | New |
|---|---|---|
| No effect: | 850 | 1170 |
| Cure: | 450 | 530 |
| Total: | 1300 | 1700 |

*Table* 3.6. Percentage responding, city and country

|  | Standard | New drug |
|---|---|---|
| City: | 16.67 | 25 |
| Country: | 50 | 60 |
| Combined: | 34.62 | 31.18 |

$\frac{530}{1700} \times 100 = 31.1\overset{8}{7}\%$. So the new drug appeared to be less effective. Indeed, a formal test (but we shall not explain the detail here) shows that, if each drug is equally effective, such a difference is unlikely to arise by chance. So there appears to be positive evidence that the standard drug does better or, equivalently, that the new drug is less effective.

Being wily, the statistician asked to see the figures for city and country separately. He used Table 3.4 to work out the percentage cures with old and new drug both for city and country. Table 3.6 gives these (you should check his calculations).

Thus, contrary to the overall impression, in both city and country the new drug gives the higher percentage cure and so does better! The statistician once again confirmed that such differences are unlikely to arise by chance if the drugs were really both equally effective.

The new drug appears to do less well *in toto* because the greater amount of testing of it was done in the city where cures are relatively harder to achieve.

### Simpson's paradox

This contradiction between the separate and combined results illustrates *Simpson's paradox*, a paradox that may arise if we combine results for city and country patients where different numbers and proportions receive each treatment in each environment and where the cure rates for each drug (Table 3.6) are markedly different in the two environments. In these circumstances direct combination may hide or even, as here, reverse the apparent effect.

The moral is that differences between proportions treated, and in the response rates, must be allowed for in assessing overall effects. After the bowling average problem in Example 3.9, you might have felt combination of data would also be appropriate here, but not so.

### Getting poor quickly

Gamblers often come unstuck with an incorrect inference about the outcome of a bet.

*Example 3.11.* I, Peter, offer to play a game with my neighbour, Paul. I will pay him £8 for each game. A game consists of tossing a fair coin until the first head appears. What Paul pays me depends on when that head appears. If it appears at the first toss, he pays me £2, leaving him a profit of £6. If it appears at the second toss, he pays me £4, leaving him a profit of £4; if at the third toss, he pays me £8, leaving us even. For each extra toss before the first head, the amount he pays me is doubled; more formally, if the first head occurs at the $k$th toss my neighbour, Paul, pays me, Peter, £$2^k$, where $2^1 = 2$, $2^2 = 2 \times 2 = 4$, $2^3 = 2 \times 2 \times 2 = 8$, etc. I pay Paul a fixed £8 at each game.

Paul admires my generosity. He knows that in the long run we get almost equal numbers of heads and tails. So *on average* about every second toss will be a head, and therefore, he argues, he will pay me on average £$2^2$ = £4 per game, whereas I pay him £8, leaving him an expected profit for £4 per game.

### Robbing Paul to pay Peter

I offer to play twenty games. Paul rubs his hands with glee, expecting to be about £4 × 20 = £80 richer after just a few minutes' fun.

Table 3.7 gives the results of each of those twenty games. In the last column a minus sign represents a loss for Paul.

Adding Paul's 'gains' in that last column, we find he has a net loss (i.e. I, Peter, have a gain) of £42. But Paul expected to gain about £80!

Convinced I'd had a lucky streak, Paul offered to play another ten games, expecting to win about £4 × 10 = £40 to break roughly even. As it turned out, in those ten games he came off even worse, losing a further £50.

I had no lucky streak. So what is wrong with Paul's reasoning? Table 3.7 pinpoints Paul's problem as those two games that went to six tosses: each cost him a staggering £56 net. He was not especially unlucky getting two such runs in twenty games. In coin tossing, runs of four, five, six or even more tails in a row before any head appears are not all that uncommon. Yet our assertion that we could expect on average about one head in two tosses was correct. It is borne out by the *number of tosses* column in Table 3.7. Adding the numbers in that column, in all twenty games the total number of tosses is forty-two, an average of $\frac{42}{20}$ = 2.1 tosses per game, a deviation from the expected value 2.0 easily accounted for by random fluctuations. Look at this another way: since all games end with the first head, there are twenty

*Table 3.7.* Robbing Paul to pay Peter

| Game no. | No. of tosses | Peter pays Paul | Paul pays Peter | Paul's net gain |
|---|---|---|---|---|
| | | £ | £ | £ |
| 1 | 1 | 8 | 2 | 6 |
| 2 | 1 | 8 | 2 | 6 |
| 3 | 3 | 8 | 8 | 0 |
| 4 | 1 | 8 | 2 | 6 |
| 5 | 1 | 8 | 2 | 6 |
| 6 | 1 | 8 | 2 | 6 |
| 7 | 1 | 8 | 2 | 6 |
| 8 | 6 | 8 | 64 | −56 |
| 9 | 1 | 8 | 2 | 6 |
| 10 | 2 | 8 | 4 | 4 |
| 11 | 2 | 8 | 4 | 4 |
| 12 | 2 | 8 | 4 | 4 |
| 13 | 3 | 8 | 8 | 0 |
| 14 | 1 | 8 | 2 | 6 |
| 15 | 1 | 8 | 2 | 6 |
| 16 | 6 | 8 | 64 | −56 |
| 17 | 1 | 8 | 2 | 6 |
| 18 | 1 | 8 | 2 | 6 |
| 19 | 3 | 8 | 8 | 0 |
| 20 | 4 | 8 | 16 | −8 |

heads and twenty-two tails in our forty-two tosses, and this does not (or should not) surprise us.

The trouble again is averages. Paul chose the *wrong* one on which to base his gain. He should have used the average payout. We learn how to calculate this in Chapter 5 (p. 85), where we find that in repeated plays of this game Paul can expect to lose all his money eventually. This is a special case of what is called the *St Petersburg paradox*.

The game looked intuitively attractive for Paul. Practice and the correct mathematical calculations show that it is not.

You may feel this sort of contradiction is only relevant to gambling and has little to do with everyday life, but not so. Problems closely related to this paradox occur in the way queues behave in practice.

### Some answers

Here are the answers to problems we posed.

*Example* 3.1 (p. 38).    $10 \times 9 \times 8 \times \cdots \times 2 \times 1 = 3\,628\,800$.

*Example* 3.2 (p. 40).    In English words $r$ is more common as the third letter. Most of us readily classify words, dictionary fashion, by the first letter. We immediately think of words like race, ran, rat, rattle, receive, etc.; less readily of words like word itself, car, cart, certainly, curtain, pertain, march, etc. Look at a page in any plain text such as a novel, and count the number of $r$'s in first and third positions. Be surprised if you find more $r$'s in the first position. Repeat the experiment for several pages.

*Example* 3.3 (p. 40).    If there are twenty-three or more people at a party, chances are better than even that at least two have the same birthday. We establish this in Chapter 4 (p. 65). Surprised? Most people guess a larger number.

*Example* 3.5 (p. 42).    Did you expect the number of days with more than 60% of the crossings from south to north to be roughly the same for each bridge? The figure exceeds 60% of all crossings for bridge A if more than twelve out of twenty crossings are south to north, whereas for bridge B it must be more than seventy-two out of 120. Now the 50:50 figures for the two bridges are, respectively, ten each way and sixty each way. In practice a discrepancy of more than two above ten (i.e. to more than twelve) occurs more often than a discrepancy of more than twelve above sixty. The *percentage* discrepancy tends to decrease as the total traffic flow increases, because, with a larger flow, discrepancies tend to even out. In practice one would probably find the number of days on which bridge A showed this 60% discrepancy to be roughly twice the number of days for bridge B.

The basic reasoning still holds if the exact numbers crossing each day (as in reality they almost certainly would) vary slightly from twenty and 120. The moral here is that in assessing regularity we must take account of the *numbers* of observations as well as *proportions* or percentages. People sometimes think proportions alone are relevant, but the numbers on which they are based are important because of this long-run tendency of fluctuations to even out. The above argument would break down if there were some linking of days on which traffic in one direction dominated on both bridges. For example, if Saturday traffic were dominantly south to north and Sunday traffic dominantly north to south on both bridges, we would need a different analysis.

*Example 3.6* (p. 42). The three outcomes are equally likely! We show in Chapter 5 that any given sequence of heads and tails in seven tosses has one chance in 128 of being observed. Many people think H–T–T–H–T–H–H is more likely than T–T–T–T–T–T–T. Their reasoning is appropriate to a different problem, namely: Are we more likely, in seven tosses, to get (i) four heads and three tails or (ii) seven tails? The answer is that four heads and three tails is more likely. This is because there are many different orderings of four heads and three tails (e.g. H–T–H–H–T–T–H, H–T–H–T–H–H–T, H–T–T–H–H–H–T, etc.). Indeed, there are thirty-five different arrangements of four heads and three tails, but only one sequence of seven tails.

### One for the road

*Example 3.12.* Think about the following problem. We discuss it in Chapter 4 (p. 65). A coin is tossed repeatedly until either the outcome *head followed by head* or *tail followed by head* appears. If the former happens first, player A wins; if the latter happens first, player B wins. Who has the better chance of winning, A or B?

### Summary

1. Intuition must be used with care; beware of anchorage effects, visual distortion, etc. (Examples 3.1 and 3.2).
2. Use all relevant information and do not be side-tracked by

irrelevant material, even if it appears to be detailed and specific (Examples 3.4 and 3.5).

3. It is important to use appropriate arithmetic procedures, especially for calculating averages or when combining data (Examples 3.8, 3.9 and 3.10).

4. Probabilistic misconceptions may lead to wrong conclusions (Examples 3.6 and 3.11).

# CHAPTER 4

# PROBABILITY AND ODDS

### Formalized common sense

We explained above that not all uncertainty problems fall within the framework of *objective probability*. To deal completely with those that do may call for expert advice; to decide if that be so, one needs to know a little about the basics of probability and statistics. In this and the two following chapters we look informally at these basics and build up the background needed for a full understanding of some of the material in Part II.

The reader meeting these concepts for the first time might be content to skim through Chapters 4–6 to get a general impression of the main ideas without the finer detail. This will suffice for a basic understanding of most of the material in Part II (except for Chapter 10). However, that understanding will be enhanced if the remaining material in Part I is eventually mastered.

Like much useful mathematics, probability theory was developed to meet practical, even if trivial, needs. It was originally developed to advise gamblers, for whom risks and benefits are often finely balanced.

### An historic example

Here is a classical probability problem. You may place an even-money bet on:

*Event A.*   In four casts of a single die there will be at least one 6.
*Event B.*   In twenty-four casts of a pair of dice there will be at least one double 6.

Are the bets equally favourable? If not, which do you prefer? The gambler Chevalier de Méré put those questions to the mathematician Blaise Pascal (1623–62). The Chevalier's intuition told him each should be equally good. This was his argument. Casting one die has six possible outcomes, scores of 1, 2, 3, 4, 5 or 6, of which one (score 6) is favourable. In four casts we expect to experience four out of six (i.e. two-thirds) of all possible outcomes. If we cast a pair of dice, since the six possible scores with the first die may be combined with any of the six for the second, there are in all $6 \times 6 = 36$ possible outcomes; only one is a double 6. In twenty-four throws we expect to experience twenty-four of the thirty-six outcomes; again the proportion is two-thirds, so each bet should be equally favourable. In practice, however, de Méré found he made a small profit on one and a small loss on the other.

Pascal recommended de Méré stick to one, predicting that, if he did, he would, with a £1 stake, in the long run show a profit of about £3.54 per 100 games.

Which did Pascal recommend and why?

## Basic objective probability

We use examples to illustrate the basic rules for handling probabilities, to show these are consistent with common sense. We often use dice, coins and cards because they are familiar. In practice we apply probabilistic ideas to real-world situations, but we have not used these for illustrations at this stage because to do so often requires explanation of non-statistical background that masks simple probability concepts.

It is easy to calculate the probability a coin falls heads; finding that for a bridge hand containing 5 spades, 4 clubs, 2 diamonds and 2 hearts is more tedious (and irrelevant to our needs in this book).

### Simple probabilities

There may be several equally likely outcomes of some action (often called a *trial* or *experiment*). If we toss a fair coin the two outcomes heads and tails are equally likely; this in essence is what we mean when we say a coin is fair. In casting a fair die the six scores 1, 2, 3, 4, 5 or 6 are each equally likely. In drawing one card from a pack of playing cards it is equally likely to be a club, spade, diamond or heart. We call any clearly defined outcome of an experiment an *event*. When we have a number of equally likely outcomes, the probability of any event of interest is given by the ratio of *the number of those outcomes favourable to that event to the total number of equally likely outcomes.*

*Example 4.1.* If I cast a die the three outcomes 1, 3 and 5 are favourable to the event *score is odd*. Thus, the probability of an odd score when a die is cast is $\frac{3}{6} = \frac{1}{2}$. We write this Pr(score is odd) $= \frac{1}{2}$, where Pr denotes 'probability'. Similarly, the probability we score 6 when a die is cast (since one out of six equally likely outcomes is favourable) is Pr(score is 6) $= \frac{1}{6}$.

More generally, if there are *n equally likely* outcomes and *r* of them favour an event *A* of interest, we say the probability of that event *A* is $r/n$ and write this Pr$(A) = r/n$.

*Example 4.2.* A box contains 122 screws; seven of them are unslotted. I grab a screw without looking. The probability it will be unslotted is $\frac{7}{122}$, since $r = 7$ and $n = 122$. The probability I select a slotted screw is $\frac{115}{122}$, since there are $122 - 7 = 115$ slotted screws.

Clearly, a probability is a positive number between 0 and 1. Events that can never occur have zero probability. The closer the probability is to 1, the *more likely* an event will happen. If something is certain to happen, it has probability 1.

### Relative frequencies

Sometimes there is no obvious set of equally likely events; we may not even be able to write down all possible outcomes. To get a measure of probability, we introduce the idea of *limiting relative frequency*.

We do not know when an individual aged sixty will die, but

insurance companies may be interested in the proportion of sixty year olds who die before age sixty-five. To estimate this, we might observe a large group of people aged sixty and note how many die before age sixty-five. If, among 1 253 219 people aged sixty, we find 132 517 die before age sixty-five, we say the relative frequency of deaths before sixty-five is

$$132\ 517/1\ 253\ 219.$$

Here is a coin tossing analogy. We toss a coin one million times. Would you be surprised to observe 500 007 heads? 499 982 heads? 500 623 heads? 397 482 heads? 27 heads?

The last two should surprise you, and make you question whether the coin is fair. We calculate the *relative frequency* of heads, denoted by Fr(H), as

$$\text{Fr(H)} = \frac{\text{number of tosses resulting in heads}}{\text{total number of tosses}}.$$

For the above observations the relative frequencies are 0.500007, 0.499982, 0.500623, 0.397482 and 0.000027, or, rounded to three decimal places, 0.500, 0.500, 0.501, 0.397 and 0.000.

The first three almost equal the probability $\frac{1}{2}$ given by the *equally likely outcomes* concept. The relative frequencies 0.397 and 0.000 are very different; they are suspect because experience tells us that *in the long run* we get about the same number of heads and tails when we toss a coin. Long-term relative frequency approaches the probability based on equally likely outcomes when this is appropriate. This suggests we use the relative frequency concept to define probabilities when we have no 'equally likely' sets to help. We say that the relative frequency tends to the probability when the number of trials is large.

Thus, for deaths of sixty year olds before age sixty-five, a total of 132 517 deaths in 1 253 219 observations gives a relative frequency

$$\text{Fr(death)} = 132\ 517/1\ 253\ 219 \approx 0.106.$$

The symbol $\approx$ means *is approximately equal to*.

This is (approximately) the probability of death in this age group *under conditions in which observations were taken*. The qualification is important. How was the group selected? Were they all workers in a particular industry? All Frenchmen? All hospital inmates? All men? All women? Occupation, nationality, health status and sex all influence life expectancy.

A probability based on relative frequency refers only to a 'population' for which the observations are in some sense a 'representative' sample (but more about this in Chapter 6). We would expect a different relative frequency among hospital patients to that among, say, civil servants.

### Calculated probabilities

We sometimes calculate a probability by combining known probabilities for simpler events using one of two laws called the *addition rule* and the *multiplication rule*.

We use $A$, $B$, $C$, ... to refer to *any* events. If only one of two events $A$ or $B$ can occur at a trial, we say the events are *mutually exclusive*. If a coin is tossed, *heads* and *tails* are mutually exclusive. We write $A$ *or* $B$ for the event where one of $A$ or $B$ occurs. The addition rule is

$$\Pr(A \text{ or } B) = \Pr(A) + \Pr(B).$$

*Example 4.3.* The probability that any car passing my house is a Ford is 0.16; the probability it is a Volkswagen is 0.08. These are *known* probabilities determined as limiting relative frequencies. What is the probability the next car is either a Ford *or* a Volkswagen?

Clearly, the events are mutually exclusive, so Pr(Ford or Volkswagen) = Pr(Ford) + Pr(Volkswagen) = 0.16 + 0.08 = 0.24.

If a number of events $A$, $B$, $C$, $D$, ... are all mutually exclusive (i.e. only one can happen at a given trial), then the addition rule extends in an obvious way. For five events, it is

$$\Pr(A \text{ or } B \text{ or } C \text{ or } D \text{ or } E)$$
$$= \Pr(A) + \Pr(B) + \Pr(C) + \Pr(D) + \Pr(E).$$

If events are not mutually exclusive, the addition rule is more complicated, but we do not need it in this book.

### Opposite events and odds

In any experiment an event $A$ either happens or it does not. The event $A$ *does not happen* is called the opposite event to $A$ (and often written $A'$). Clearly, $A$ and $A'$ are mutually exclusive and cover all possibilities. Thus, the event $A$ *or* $A'$ is sure to happen and has probability 1. The ratio $\Pr(A'):\Pr(A)$ defines the *odds against* the event $A$.

*Example 4.4.* A die is cast. If $A$ is the event *score is 6*, then $A'$ is the event *score not 6* (i.e. it is 1, 2, 3, 4 or 5). We know $\Pr(A) = \frac{1}{6}$; the 'equally likely events' concept gives $\Pr(A') = \frac{5}{6}$, so the odds against a six are 5 to 1.

We used the idea of odds intuitively in Example 3.4. Since $A$ and $A'$ are mutually exclusive,

$$\Pr(A \text{ or } A') = \Pr(A) + \Pr(A').$$

Since $A$ and $A'$ are exhaustive (i.e. one must occur),

$$\Pr(A \text{ or } A') = 1.$$

Combining these, we have $\Pr(A) + \Pr(A') = 1$, whence

$$\Pr(A) = 1 - \Pr(A').$$

This may seem trivial, even pedantic, but it is useful because sometimes it is easier to calculate $\Pr(A')$ than $\Pr(A)$; we then obtain the latter by subtraction from 1. We demonstrate this in Example 4.10.

### Conditional probability

On pp. 57–58 we considered the probability of a person aged sixty dying before age sixty-five, conditional upon their being a hospital inmate, civil servant, etc. We formalize the probability of an event *conditional* upon some other event.

*Example 4.5.* I draw a card from a well-shuffled pack of playing cards. Since thirteen of the fifty-two cards are hearts, the probability I draw a heart is $\frac{13}{52} = \frac{1}{4}$. The probability it is not a heart (the opposite event to a heart) is $1 - \frac{1}{4} = \frac{3}{4}$. I now draw a second card (without replacing the first). What is the probability it is a heart?

The answer depends on whether or not the first card was a heart. If the first *was* a heart, then twelve of the remaining fifty-one cards are hearts. All have an equal chance of being drawn, so Pr(second card a heart) $= \frac{12}{51}$. If the first card was *not* a heart, we select the second from fifty-one cards of which thirteen are hearts; whence Pr(second card a heart) $= \frac{13}{51}$. These are called *conditional probabilities*. The probability $\frac{12}{51}$ is the probability the second card is a heart *conditional upon the first being a heart*; the probability $\frac{13}{51}$ is the probability the second card is a heart *conditional upon the first not being a heart*.

For any two events *A* and *B*, the *probability of B conditional on A* is the probability that *B* occurs when *A* has already occurred. The shorthand for this is $\Pr(B|A)$; the vertical bar is read as *conditional on*.

If *A* is the event *first card drawn is a heart* and *B* the event *second card is a heart*, then $\Pr(B|A) = \frac{12}{51}$. If *A'* is the opposite event to *A* (i.e. *first card not a heart*) then $\Pr(B|A') = \frac{13}{51}$.

## The multiplication rule

If *A* and *B* are two events, Pr(*A* and *B*) is the probability both occur. This is calculated by the *multiplication rule*, which states:

$$\Pr(A \text{ and } B) = \Pr(A) \times \Pr(B|A) = \Pr(B) \times \Pr(A|B).$$

*Example 4.6.* From Example 4.5, we see that if *A* and *B* are the events *first/second card is a heart*, then the probability we draw two hearts is

$$\Pr(A \text{ and } B) = \Pr(A) \times \Pr(B|A) = \frac{13}{52} \times \frac{12}{51} = \frac{1}{17}.$$

### Independent events

Two events are *independent* if $\Pr(B|A) = \Pr(B)$. This means the probability of $B$ is the same *whether or not A* also occurs. The multiplication rule then takes the simpler form

$$\Pr(A \text{ and } B) = \Pr(A) \times \Pr(B)$$

and this implies $\Pr(A|B) = \Pr(A)$.

*Example 4.7.*    If we cast a die twice (or cast two dice at the same time), outcomes are independent; for example, the probability of a 6 (or any other score) at any cast is always $\frac{1}{6}$ irrespective of the score at the other cast. Thus, the probability of a double 6 when a die is cast twice is $\frac{1}{6} \times \frac{1}{6} = \frac{1}{36}$.

We can extend the multiplication rule to several events. In the general form we consider each event conditional upon all previous relevant events having occurred, and multiply the results. For example, if we select three cards from a pack without replacement, we already know that the probability the first and second are hearts are, respectively, $\frac{13}{52}$ and $\frac{12}{51}$. We select the third card from fifty of which eleven are hearts (because two hearts have already been selected). Thus, the probability we select three hearts is $\frac{13}{52} \times \frac{12}{51} \times \frac{11}{50} = \frac{11}{850}$.

When events are independent, we simply multiply the probabilities of each. For example, in three throws of a fair die the probability of three 6's is $\frac{1}{6} \times \frac{1}{6} \times \frac{1}{6} = (\frac{1}{6})^3 = \frac{1}{216}$. An obvious extension may be made to more than three events.

*Example 4.8.*    Three cards are drawn from a pack with replacement (i.e. after each is drawn it is replaced and the pack well shuffled). What is the probability all cards drawn are hearts?

For each of the three drawings $\Pr(\text{heart}) = \frac{1}{4}$, since we draw from a full pack each time. Thus,

$$\Pr(\text{all 3 cards hearts}) = \frac{1}{4} \times \frac{1}{4} \times \frac{1}{4} = (\frac{1}{4})^3 = \frac{1}{64}.$$

### Built-in safety

*Example 4.9.* For safety, vital aircraft or machine components are often duplicated or triplicated. If we have $n$ components and the probability $p$ of failure in any one is 0.001, and failures are independent, the probability, given by the multiplication rule, that all $n$ fail is $p^n = 0.001^n$.

For $n = 1, 2$ and 3, the probabilities are 0.001, 0.000 001 and 0.000 000 001. If the machine only breaks down when all components fail, note the rapidly increasing safety margin (reflected in the much lower probability of a breakdown) as the number of back-up components increases. Risk is reduced dramatically by duplication or triplication of a vital component.

We have assumed that the failures were independent. This would not be realistic if failure of one component put additional strain on the remaining ones. If this be so, modified calculations of the probability that all fail would be required.

A more complicated type of built-in safety mechanism is used for systems enabling aircraft to land in near-zero visibility. Here, the problem is that part of the system may malfunction without it being obvious that it is doing so. For this reason, before approving one such system, the U.K. Civil Aviation Authority (CAA) insisted, not only on a low probability of failure, but also on protection against the difficulty of recognizing a malfunction. Duplication of vital elements would not suffice, for, if two units did not agree, it might not be evident which was incorrect. The CAA insisted on triplication. If one unit disagreed with two others, it would be reasonable to assume that the odd one was malfunctioning and to accept information from the two that agreed. The empirical evidence is strong that there is a negligible probability of identical units simultaneously malfunctioning in a way that would produce exactly the same erroneous information. If all three give different information, a landing is immediately aborted!

### De Méré's gamble

*Example 4.10.*   A die is cast four times. What is the probability of at least one 6?

Exactly one 6, exactly two 6's, exactly three 6's and exactly four 6's are mutually exclusive ways of getting at least one 6. We could apply an extended addition rule but this is tedious. It is easier to note that 'at least one 6' is the opposite event to 'no 6'. At each of the four throws, the probability of *not* scoring a 6 is clearly $\frac{5}{6}$, for it is the *opposite* event to scoring a 6, which has probability $\frac{1}{6}$. Outcomes at each throw are independent, so, using the extended multiplication rule, it follows that the probability of no 6 at all four throws is $(\frac{5}{6})^4$. Hence,

$$\begin{aligned}
\Pr(\text{at least one 6}) &= 1 - \Pr(\text{no 6}) \\
&= 1 - (\tfrac{5}{6})^4 \\
&= 1 - \tfrac{625}{1296} \\
&\approx 0.5177.
\end{aligned}$$

*Example 4.11.*   De Méré's alternative was an even-money bet on at least one double 6 in twenty-four throws of a pair of dice.

In Example 4.7 we found the probability of a double 6 is $\frac{1}{36}$. Thus, the probability of no double 6 at one throw is $1 - \frac{1}{36} = \frac{35}{36}$. Generalizing the multiplication rule to twenty-four throws, we have

$$\Pr(\text{no double 6}) = (\tfrac{35}{36})^{24} = 0.5086,$$

a result easily got using a pocket calculator with a $y^x$ key, or by using logarithms. Now

$$\Pr(\text{at least one double 6}) = 1 - \Pr(\text{no double 6}),$$

these being opposite events. So, we have

$$\begin{aligned}
\Pr(\text{at least one double 6}) &= 1 - (\tfrac{35}{36})^{24} \\
&= 1 - 0.5086 \\
&= 0.4914.
\end{aligned}$$

Note that, by the definition of *odds against* (p. 60), the odds against a win in Example 4.11 are 0.5086 to 0.4914, or 1.0350 to 1.

*Odds against* greater than *evens* (i.e. 1 to 1) mean an even-money bet is *unfavourable* and in the long run the gambler will lose. Odds against less than 1, as we have in Example 4.10, are favourable. We discuss in Example 5.8 de Méré's expected gains with each wager.

Can you now see the flaw in de Méré's reasoning (p. 56)?

### That one for the road

In Example 3.12 we tossed a coin until we got either two consecutive heads, in which case player A won, or a tail followed by a head, in which case B won. It is tempting to say each is equally likely to win since the multiplication rule shows the sequence HH or TH are equally likely, each having probability $\frac{1}{4}$; but not so. A wins if the first two tosses are HH. For any other pair of tosses at the start (i.e. TH, HT or TT), B will automatically win, since in all these cases it is easily seen that a sequence TH *must* occur before HH. Indeed, in the first of these, case TH occurs immediately. Since the probability of starting HH is $\frac{1}{4}$, and of getting some other starter sequence is $\frac{3}{4}$, the odds are 3 to 1 against A winning. Were your hunches right about this?

### The birthday problem

In Example 3.3 we asserted (p. 52) that, if there were at least twenty-three people (none twins) at a party, there was a better than even chance that at least two would have the same birthday.

We assume anyone's birthday is equally likely to be on any day of the year. This is not exactly true, but variations are slight with, for example, fewer births on Christmas Day, thanks to artificial means of inducing a birth. The extra day in a leap year is another complication, but such factors have virtually no effect on the argument used below.

For any two people at the party, the first person's birthday may be *any* day; the second person has a different birthday if it occurs on one of the remaining 364 days. So the probability is $\frac{364}{365}$ that he or she will have a different birthday. A third person's birthday will be different from each of the first two if it occurs on any of the remaining 363 days; this has probability $\frac{363}{365}$. Thus, the probability that *both* the second and third person have birthdays different to each other and to the first is, by the multiplication rule, $\frac{364}{365} \times \frac{363}{365}$. The probability a fourth person has yet a different birth date is $\frac{362}{365}$, a fifth $\frac{361}{365}$, a sixth $\frac{360}{365}$, and so on until we come to the twenty-third person, for whom it will be $\frac{343}{365}$. By an extended multiplication rule, the probability all twenty-three people have different birthdays is

$$\frac{364}{365} \times \frac{363}{365} \times \frac{362}{365} \times \cdots \times \frac{345}{365} \times \frac{344}{365} \times \frac{343}{365}.$$

Watch out for rounding errors if you use a pocket calculator. To avoid these, first divide 364 by 365, then multiply by 363 and divide by 365, multiply by 362 and divide by 365, continuing to the stage where you multiply by 344 and divide by 365. Your answer then (to three decimal places) should be 0.524. Now multiply by 343 and divide by 365 to get the final answer: it is 0.493. This is the probability no two of twenty-three people have the same birthday. The opposite event, *at least two have a common birthday*, has probability $1 - 0.493 = 0.507$, greater than a half. If there are only twenty-two people, the corresponding probability is $1 - 0.524 = 0.476$, less than a half. Clearly, for each additional person, the probability of at least one coincidence increases. The probabilities of at least one coincidence for parties of various sizes are given in Table 4.1. Regarding probability as limiting relative frequency, we can see that these results imply that, if we went to a lot of parties, each with fifty people present, then at about ninety-seven out of every hundred we would find at least one coincidence of birthdays. Did your intuition suggest so high a probability?

## Probability and terminology

Chapter 3 exposed some dangers of intuition when operating with data. Intuitive ideas about probabilities may also be wrong. We saw this on p. 50 for the game *Robbing Paul to Pay Peter* and

*Table 4.1.* Probabilities of at least two
birthdays coincident for parties of
various sizes

| Number present | Probability |
|----------------|-------------|
| 22 | 0.476 |
| 23 | 0.507 |
| 30 | 0.706 |
| 40 | 0.891 |
| 50 | 0.970 |

again in de Méré's reasoning (p. 56) why two dice games should be equally favourable.

Verbal misunderstandings about probabilities are also common. To assess pupils' skills at assessing and comparing probabilities, Green (1984) posed a number of multiple choice questions to children of various ages and abilities. Two of his questions are given below. In each case before reading further decide which answer you think is correct.

*Example 4.12.* In an experiment twelve coins are tossed in the air together, and land on a table. If the experiment is repeated a lot of times, which one of the following results will happen most often?

(A) Two heads and ten tails
(B) Five heads and seven tails
(C) Six heads and six tails
(D) Seven heads and five tails
(E) All have the same chance.

For 2930 children aged between eleven and sixteen, Green found that almost irrespective of age about 70% thought (E) was correct. The correct answer is (C). This follows from the binomial distribution (Chapter 5, p. 78). The problem was one of language: interrogation showed many pupils believed 'All have the same chance' meant the same as 'All have some chance', rather than being equivalent to 'All are equally likely'. Only about 12% of younger children and 20% of older children gave the correct answer.

*Example 4.13.* For the following phrases, tick all those which you think mean *exactly* the same as 'has a 50–50 chance of happening'.

(A) It may happen or it may not
(B) It has an even chance of happening
(C) It will happen fifty times out of fifty
(D) It can happen sometimes
(E) It has an equal chance of happening or not happening
(F) It is very unlikely to happen.

Most pupils rightly selected (B) and (E) as equivalents, but often added other choices. Two-thirds of the younger children and three-quarters of the older ones thought (A) an exact equivalent. They did not seem to realize that (A) would still be true if the event only had, say, one chance in five of happening.

Precise phraseology is important for a clear understanding of probability concepts.

### One-off events

Objective probabilities imply the idea of repeatable experiments: tossing a coin, casting a die, observing survival rates for many sixty year olds. We cannot define in this way the probability that the Prime Minister of Great Britain in the year 2000 will be a woman, for that is a 'one-off' event. It is clearly wrong to argue that the probability is about one-half on the grounds that approximately half the people in Britain are women. Not everybody is equally likely to become Prime Minister in the year 2000!

Yet most of us have an opinion (on which we might be prepared to bet) about how likely it is that the Prime Minister in the year 2000 will be a woman. We might express this as a *personal* or *subjective* probability, or by equivalent odds. A rational person does this using information about British politics. For example, it is unusual for anyone to become Prime Minister (P.M.) without first serving some years in Parliament. After the election in June 1987, there were forty-one women in a Parliament of 650 members.

Not all M.P.s are potential future P.M.s; indeed few are. They must be people who combine ambition with certain abilities and party affiliations. It is quite possible that the P.M. for the year 2000 is already an M.P., and experts in politics could name a few likely future P.M.s. These might include one or two women, rather more men. You may believe that women will play an increasingly active role in politics, so that in future about one P.M. in three is likely to be a woman; or you may think only one in five, or one in ten. Weighing these and other relevant factors (e.g. Which party is likely to be in power in the year 2000? Does it have many women M.P.s? And so on), you could arrive at what you think would be fair betting odds against a woman P.M. in the year 2000. My guess at the moment (January 1988) is that many people would probably feel that odds somewhere between 2 to 1 and 10 to 1 against might not be unreasonable. You may not agree. Whether you do might depend on when you read this. In the 1990s you would possibly arrive at different reasonable odds in the light of new information. There may by then be more women M.P.s who are predicted to become P.M.s, so shortening the odds to 2 to 1 against or even a good even-money bet. If, on the other hand, by 1992 there were only two female M.P.s, each members of minority parties, you might then consider 100 to 1 against to be reasonable odds.

## Personal probabilities and risks

In many situations such as 'one-off' risks where objective probabilities are not available, we intuitively arrive at personal or subjective probabilities (or corresponding odds) that we regard as reasonable. These differ for individuals, particularly when useful and relevant information is scarce. Those who fear an accident at a nuclear power station that may destroy civilization assign a higher personal probability to that event than does someone who considers it remote.

Rational people are usually in closer agreement about subjective probabilities if they share a large store of correct and relevant information and have reliable guidance on how to interpret it. An ultimate hope is that all rational people, given the same sufficient evidence, would arrive at the same probability for a 'one-off' event. For example, if a lot of people are asked 'What are the odds that it will rain within the next hour?', many will be in close agreement, because they know how to interpret the signs for short-term weather prospects; there may be more diversity if they are asked to quote odds on rain two days hence. However, in parts of the world where there are clearly defined wet and dry seasons, agreement is likely to be high even when forecasting many days ahead.

If there is little relevant information, one piece of critical evidence may profoundly affect personal probabilities. Many people revised their probabilities for a civilization-destroying nuclear catastrophe after Chernobyl and the associated publicity.

The addition and multiplication rules apply to subjective probabilities if we are consistent in our allocation. By consistency we mean, for example, that if we assign a subjective probability of 0.2 to the P.M. in the year 2000 being a woman, we must assign a probability of 0.8 to the opposite event (a man).

## Utility

For de Méré's gambles, odds for or against winning indicated a rational choice of game. We often face situations where we, quite willingly, take an action that leaves us with an expected loss (negative gain). Punters know that bookmakers select their odds, or probabilities, to ensure that they, not the punter, experience an expected gain, implying an expected loss for punters as a whole. Why, then, does the man in the street like a flutter on the Derby or the Grand National? Clearly, he is not concerned solely with long-term gain if he bets often.

### Lotteries

With lotteries the punter's reasoning is even clearer. For most major lotteries, we may calculate *objective* probabilities of winning various amounts and most investors simply lose their stake money. The organizer of the lottery takes in more than he pays out (or he will soon be out of business). The attraction of a lottery is the small probability of winning that big prize. We are prepared to lose (almost certainly) a small stake – perhaps £1.00 each week – for the small chance of becoming a millionaire. The £1.00 stake might of course be spent on something else, but, if it is only a small part of our assets, we may prefer to invest it in the lottery to spending it on a tot of whisky or buying a magazine. We say the £1.00 stake has a small *utility value* (we are somewhat indifferent about how we spend it); the elusive million pound prize has a high utility value. We are prepared to forego a weekly magazine or a drink for the outside chance of becoming a millionaire. Most of us will not achieve that goal; a lucky few will.

What we learn from this simple example is that expectation of a virtually certain long-term gain or loss is not enough to explain our behaviour. Even those who think betting foolish do something very like it when they buy insurance.

### *Fire*

The (objective) probability, based on limiting relative frequencies, that a conventional suburban house will burn down in any year is about 1 in 1000. You might insure your house for £60 000 for an annual premium of £100. This means in any year you *lose* £100 with probability $\frac{999}{1000}$ or retain the asset value of £60 000 (less your £100 premium) with probability $\frac{1}{1000}$

In the long run the insurance company expects, for risks like yours, to pay out one claim of £60 000 on every 1000 premiums of £100; that is, it collects £100 000 in premiums for every £60 000 paid out, retaining £40 000 per 1000 premiums (or £40 per premium) for its expenses and profits. We say a policyholder's *expected loss* is £40 per annum. In most years he loses his £100. If his house is burnt down he is recompensed with a £60 000 payout.

You insure your house because of its utility value; it is, for most of us, a large portion of our assets. It is more serious to lose £60 000, a substantial part of our assets, at one blow than it is to steadily lose £100 each year if we make no claim.

Most of us prefer a highly likely loss of £100 in each of perhaps some forty years of home ownership to a small chance of losing £60 000 of assets at a stroke.

Utility depends on circumstances. The £100 for the premium is probably a small proportion of our annual income; by paying it, we may be foregoing a dinner for four at a top-class restaurant or a weekend break. For those of us on a modest income, our whole lifestyle will change dramatically (and unpleasantly) if our house is uninsured and destroyed.

For a few rich individuals, a £60 000 house is a small portion of their total assets, and their utilities may be different to those of a typical wage or salary earner. A millionaire who has a second home worth £60 000 (his first home would probably be worth much more) might not bother to insure it. The odds are it won't burn down; if it does, he can afford to replace it by selling a few shares.

You or I may think of a £60 000 asset in much the same way a millionaire thinks of a £600 000 asset; each may be about the same fraction of our respective total assets. For the millionaire, £60 000 then may have about $\frac{1}{10}$ of the utility value it has for you or me. On the other hand, the £100 that might be used for an insurance premium has perhaps a very similar utility for the average man (if he is not too poor) and for the millionaire in the sense that it will still buy a good dinner for four or a weekend break. We tend to assess utility rather differently for income than we do for capital assets.

For a large property company owning, say, 800 houses valued at £60 000 each, the utility situation may not be unlike that for the millionaire, but for different reasons. To insure all 800 at a premium of £100 each would cost £80 000 per annum. The expected loss for the insured (or gain for the insurer) is now 800 times as much as that for an individual with only one house (i.e. £40 × 800 = £32 000), so the property company might decide not to insure at all and keep the £80 000 required for premiums, a course often referred to as *carrying one's own insurance*. The company's utilities differ from those of an individual. Of course, a property company would take other factors into account before making a final decision. If the houses were close together on one estate, it must not forget the possibility of one disastrous fire destroying a whole block of houses. Even so, if it decides to insure, it will almost certainly be able to negotiate a better bulk premium.

Utility plays an important role in determining attitudes to uncertainty. Just as we have developed a formal probability theory, there is also a formal theory for making decisions based on utilities. This is described in easily understandable terms by Moore (1983) in *The Business of Risk* and in slightly more general terms by Barnett (1982) in *Comparative Statistical Inference*.

## Summary

1. If the event $A$ occurs in $r$ out of $n$ *equally likely* outcomes, then $\Pr(A) = r/n$. Probabilities may also be defined as limiting relative frequencies.

2. *Addition rule*. If $A$ and $B$ are mutually exclusive (i.e. only one can occur at a given trial), then $\Pr(A \text{ or } B) = \Pr(A) + \Pr(B)$. If $A'$ is the opposite event to $A$ (i.e. $A'$ is the event $A$ *does not happen*), then $\Pr(A') = 1 - \Pr(A)$.

3. The probability that $B$ occurs, given that $A$ has also occurred, is called the *probability of B conditional upon A*, and is written $\Pr(B|A)$.

4. *Multiplication rule*. This is

$$\Pr(A \text{ and } B) = \Pr(A) \times \Pr(B|A) = \Pr(B) \times \Pr(A|B).$$

If $A$ and $B$ are independent, then $\Pr(A \text{ and } B) = \Pr(A) \times \Pr(B)$.

5. Personal and subjective probabilities may be assigned to one-off events. These may differ from person to person, but differences among rational people are usually small if each is based on the same large store of relevant information.

6. *Utility* may influence some actions more strongly than expected loss or gain, but utilities vary between individuals and between groups.

# CHAPTER 5

## APPLIED PROBABILITY

### Permutations and combinations

In applications of probability we often have to count the number of ways we can do things, in particular the numbers of ways of selecting objects from a larger group. The order of selection may or may not be important.

*Example* 5.1. There are twelve horses in a race. How many different forecasts can be made of horses finishing first, second and third? About 50? About 100? About 500? About 1000? About 10 000?

The answer is 1320. With time and patience all can be written down, but, if we only want to know how many, we need not list each one. We have twelve choices for the horse placed first, leaving eleven choices for second; thus, there are $12 \times 11 = 132$ ways of writing down the first two placegetters. For each of these 132 ways, there are ten ways of filling third place (one of the ten horses not already chosen as first or second); this means there are $132 \times 10 = 1320$ ways to select all three placegetters.

The number of ways of selecting $r$ from $n$ distinguishable objects when order is important is called the *number of permutations of $r$ objects from $n$* and is written $^nP_r$.

Generalizing arguments in Example 5.1, we have $n$ choices for first place, $n - 1$ for the second, which combine to give $n \times (n - 1)$ ways of selecting the first two objects. The third may now be any of the remaining $n - 2$, giving $n \times (n - 1) \times (n - 2)$

*Table 5.1.* Possible orderings of four different items

| ABCD | ABDC | ACBD | ACDB | ADBC | ADCB |
|------|------|------|------|------|------|
| BACD | BADC | BCAD | BCDA | BDAC | BDCA |
| CABD | CADB | CBAD | CBDA | CDAB | CDBA |
| DABC | DACB | DBAC | DBCA | DCAB | DCBA |

ways of selecting the first three. The fourth may now be selected in $n - 3$ ways, giving

$$^n P_4 = n \times (n - 1) \times (n - 2) \times (n - 3).$$

Continuing, we find that the $r$th object may be selected in $n - r + 1$ ways, whence

$$^n P_r = n \times (n - 1) \times (n - 2) \times \cdots \times (n - r + 1). \quad (5.1)$$

If we put $r = n$ in (5.1), we get

$$^n P_n = n \times (n - 1) \times (n - 2) \times \cdots \times 3 \times 2 \times 1,$$

that is, the product of all integers between 1 and $n$. This product is given the special name *factorial n* and denoted by the symbol $n!$ ($n$ followed by an exclamation mark). Factorial $n$ gives the number of different ways we can arrange $n$ items in order. In Example 3.1 we met 10!; its value is more than three million. Some scientific calculators have a key for factorial $n$, usually labelled **n!** or **x!**.

Given four items A, B, C, D there are $4 \times 3 \times 2 \times 1 = 24$ orderings. These are listed in Table 5.1.

*Example 5.2.* How many permutations of four items from five are possible? Here $n = 5$ and $r = 4$, so $^5 P_4 = 5 \times 4 \times 3 \times 2 = 120$.

Labelling the items A, B, C, D, E, twenty-four of these permutations (i.e. all those without the letter E) are given in Table 5.1. We get another twenty-four by replacing A by E, a further twenty-four by replacing B by E, the next twenty-four by replacing C by E, and the final twenty-four by replacing D by E.

## Does order matter?

For prizes, or betting on horses, place order is important. In a bridge or poker hand only the cards matter, not the order in which they are dealt, or if three singers are selected from fifteen to appear as a trio at a concert and all get the same publicity and pay, order of selection is unimportant.

Table 5.1 listed the 24 *ordered* selections of the one group A, B, C, D from five items A, B, C, D, E. If we are not interested in order, the important point is that all involve the *same* four symbols A, B, C, D. From five objects, we can select only four other sets of four, viz. E, B, C, D or A, E, C, D or A, B, E, D or A, B, C, E.

The number of ways of selecting $r$ items from $n$ when order is ignored is called *the number of combinations of r items from n* and written $^nC_r$, or sometimes $\binom{n}{r}$. Clearly, the items in each *different* combination of $r$ may be arranged among themselves in $r!$ ways, each giving a different *permutation*. Thus,

$$^nC_r \times r! = {}^nP_r, \tag{5.2}$$

implying

$$^nC_r = {}^nP_r/r!. \tag{5.3}$$

This is useful for calculating $^nC_r$. It is not difficult algebra to show that

$$^nC_r = \frac{n!}{r!(n-r)!}. \tag{5.4}$$

## Card games

*Example 5.3.*    How many different poker hands (five cards) can be dealt from an ordinary pack of fifty-two cards?

This is a combinations problem since it is irrelevant in which order the cards are dealt. Using (5.4), we find that the number of hands is $^{52}C_5 = 52!/(47! \times 5!)$. After cancellation between numerator and denominator, this simplifies to

$$^{52}C_5 = \frac{52}{5} \times \frac{51}{4} \times \frac{50}{3} \times \frac{49}{2} \times \frac{48}{1} = 2\,598\,960.$$

This also follows directly from formula (5.3). There are more than two-and-a-half million different poker hands. Are you surprised? Each hand is equally likely. If you have a calculator with a **n!** key, a sequence like 52! ÷ 47! ÷ 5! = may be used.

Replacing $r$ by $n - r$ in (5.4) does not change the right-hand side, so

$$^nC_r = {^nC_{n-r}}. \tag{5.5}$$

Relationship (5.5) saves a lot of calculation.

If $r = 0$, both (5.3) and (5.4) contain 0! in the denominator. To give this a meaning, we define 0! to be unity. Then we see from (5.4) that $^nC_0 = 1$. This is sensible, for $^nC_0$ is the number of ways of selecting no items from $n$. There is one way to do this – leave them all out! There is also only one way of selecting $n$ things from $n$ – take the lot!

### *Committees*

In Example 3.7 we claimed that 220 different committees of size 3 can be selected from twelve people. This follows from formula (5.3), since $^{12}C_3 = (12 \times 11 \times 10)/(1 \times 2 \times 3) = 220$. Formula (5.5) implies $^{12}C_9 = {^{12}C_3} = 220$ also. The number of committees of size 6 is $^{12}C_6 = 924$. You should calculate also the numbers of committees of size 4, 5, 7 and 8 to verify our assertion in Example 3.7 that there are more committees of 6 than any other size.

### Bernoulli trials

A Bernoulli trial is an experiment that has *only* two mutually exclusive outcomes with respective probabilities $p$ and $q$. Clearly, $q = 1 - p$ (opposite events) and $p + q = 1$.

Tossing a coin is a Bernoulli trial with outcomes *heads* or *tails*, if the coin is fair, or unbiased, $p = q = \frac{1}{2}$, where $p = \Pr(\text{heads})$. Observing the sex of a new born child is a Bernoulli trial with, approximately, $p = \Pr(\text{male}) = \frac{1}{2}$. If a die is cast and we are only interested in whether or not we score a 6, this is a Bernoulli trial with $p = \Pr(6) = \frac{1}{6}$ and $q = \Pr(\text{not } 6) = \frac{5}{6}$. If a batch of a hundred machine-produced items contains ninety-six good and four faulty items, selecting an item and noting whether it is good or faulty is a Bernoulli trial with $p = \Pr(\text{good}) = 0.96$. If leaves on a plant may be infected with probability $p$, or not infected with probability $q$, selecting a leaf and noting whether or not it is infected is a Bernoulli trial.

### Repeated trials

Suppose we perform some fixed number $n$ of independent Bernoulli trials, each with the same $p$ and $q$. What can we say about the number of times the event with probability $p$ occurs? For example, in a family of seven we may ask what are the probabilities that

  (i) exactly three are boys?
 (ii) there are more boys than girls?
(iii) all seven children are boys?

Different physical situations may lead to the same statistical model. The probabilities in the following problems will be the same as those just sought for children.

A coin is tossed seven times. What are the probabilities of

  (i) exactly three heads?
 (ii) more heads than tails?
(iii) all seven tosses are heads?

The common link is that we have the *same number* ($n = 7$) of Bernoulli trials and at each, for the event of interest, $p = \frac{1}{2}$.

### The binomial distribution

In $n$ independent Bernoulli trials with the same $p$ and $q$, the event with probability $p$ may occur any number of times between 0 and $n$. The precise number will vary each time we perform this experiment. The number of occurrences is called a *random variable* because its value is determined by a chance mechanism. We often use $X$ as a shorthand for a random variable. It is convenient to call the outcome with probability $p$ a *success*, often abbreviated to S. The other outcome, with probability $q$, is denoted by F (for failure). Using the term *success* for convenience does not imply, for instance, that a male birth is a success and a female birth a failure in the everyday meaning of these words.

We seek an expression for $\Pr(X = r)$ when $r$ takes any integer value between 0 and $n$. To get this, we use the multiplication and addition rules for probabilities and the concept of combinations.

Suppose the number of trials $n$ is 7 and the number of successes $r$ is 3 (implying $7 - 3 = 4$ failures). First, we calculate the probability of three successes and four failures in a *particular* order (e.g. S–F–F–S–F–S–F). The outcomes of all seven trials are independent, so the probability of this particular sequence is the product of the probabilities of the events at each trial. Writing them in the order above, it is $p \times q \times q \times p \times q \times p \times q = p^3q^4$.

It is clear that, if we have three successes and four failures in some other order, the probability of getting the result in *that different order* is again $p^3q^4$.

Each different ordering represents a *mutually exclusive* way of getting three successes and four failures, so the addition rule is relevant. Finding the number of different orders is a combinations problem. Effectively, we have seven slots and choose three for the letter S; the remaining four then automatically receive the letter F. In other words, we must select three from seven positions for the letter S: this can be done in $^7C_3$ ways. Using the addition rule, we finally obtain

$$\Pr(X = 3) = {}^7C_3 p^3 q^4.$$

If, instead of seven, we have $n$ trials, and, instead of three, we have $r$ successes, the above argument generalizes to give

$$\Pr(X = r) = {}^nC_r p^r q^{n-r}, \tag{5.6}$$

*Table 5.2.* Probability distribution of numbers of boys in families of seven

| Number of boys: | 0 | 1 | 2 | 3 | 4 | 5 | 6 | 7 |
|---|---|---|---|---|---|---|---|---|
| Probability: | $\frac{1}{128}$ | $\frac{7}{128}$ | $\frac{21}{128}$ | $\frac{35}{128}$ | $\frac{35}{128}$ | $\frac{21}{128}$ | $\frac{7}{128}$ | $\frac{1}{128}$ |

where $r$ may be any integer between 0 and $n$ inclusive.

If we know the probability associated with each possible value of $X$, we say we know the probability distribution of $X$. If the probabilities are given by (5.6), we call the probability distribution a *binomial distribution*. If you know a theorem in algebra called the binomial theorem you will see why, but if not, you only need to remember the name.

The binomial distribution provides a sophisticated way of working out the probability of $r$ boys in a family of $n$.

*Example 5.4.* Putting $n = 7$ and $p = q = \frac{1}{2}$ in (5.6) answers the questions posed on p. 78. It is clear from (5.6) that the probability of exactly three boys (implying four girls also) in a family of seven is

$$\Pr(X = 3) = {}^{7}C_3(\tfrac{1}{2})^3 \times (\tfrac{1}{2})^4 = \tfrac{35}{128}.$$

To answer the other questions on p. 78, it is convenient, though not essential, to write down the complete binomial distribution using (5.6). If $r$ denotes the number of boys, we have

$$\Pr(X = r) = {}^{7}C_r(\tfrac{1}{2})^r \times (\tfrac{1}{2})^{7-r}.$$

This simplifies to

$$\Pr(X = r) = {}^{7}C_r(\tfrac{1}{2})^7, \tag{5.7}$$

where we have used the algebraic rule for indices that $a^x \times a^y = a^{x+y}$.

It is easily verified that substitution of appropriate values in (5.7) gives $\Pr(X = 0) = (\tfrac{1}{2})^7 = \tfrac{1}{128}$, $\Pr(X = 1) = \tfrac{7}{128}$, $\Pr(X = 2) = \tfrac{21}{128}$, etc. Table 5.2 gives probabilities for all value of $X$.

Now we can easily answer the other questions. There will be

more boys than girls if there are 4, 5, 6 or 7 boys. These are mutually exclusive outcomes, so we add probabilities:

$$\text{Pr(more boys)} = \text{Pr}(X = 4) + \text{Pr}(X = 5)$$
$$+ \text{Pr}(X = 6) + \text{Pr}(X = 7)$$
$$= \frac{35 + 21 + 7 + 1}{128} = \tfrac{64}{128} = \tfrac{1}{2}.$$

We could have got this more easily. Clearly, a family of seven will either have more girls *or* more boys and if the probabilities of each birth being male or female are identical, we are equally likely to have either outcome. No other outcome is possible, so each has probability one-half.

What about a family of six? It is still equally likely there will be more boys than girls or more girls than boys, but the probability of each of these outcomes is no longer one-half; clearly, now, there is a non-zero probability of having the same number of boys as girls. What is this probability?

Finally, from Table 5.2 we see that the probability that for families of seven, all are boys is $\tfrac{1}{128}$. These probabilities apply also to numbers of heads (or tails) in seven tosses of a coin; this is relevant in Example 3.6.

### Dice

*Example* 5.5.   I cast a die five times and win a prize if I get a 5 or 6 on three or more of the five casts. What is the probability I win? The probability of a five or six is $\tfrac{1}{3}$ at each cast. Call this a success. The number of successes has a binomial distribution with $n = 5$ and $p = \tfrac{1}{3}$. If $X$ is the random variable *number of successes*, then

$$\text{Pr}(X = r) = {}^5C_r(\tfrac{1}{3})^r(\tfrac{2}{3})^{5-r}.$$

Substituting the values $r = 0, 1, 2, \ldots$, we get the probability for each number of successes given in Table 5.3.

From this table, we find the probability of three or four or five successes by adding $\text{Pr}(X = 3) + \text{Pr}(X = 4) + \text{Pr}(X = 5) = 0.2099$.

*Table 5.3.* Probability of *r* scores of 5 or 6 in five casts of a die

| *r*: | 0 | 1 | 2 | 3 | 4 | 5 |
|------|------|------|------|------|------|------|
| Pr($X = r$): | 0.1317 | 0.3292 | 0.3292 | 0.1646 | 0.0412 | 0.0041 |

## Graphs or tables

If you find it easier to visualize patterns from a graph, a convenient picture of binomial probabilities is obtained by centring rectangles of unit breadth at $r = 0, 1, 2, \ldots, n$, each with height equal to the corresponding probability.

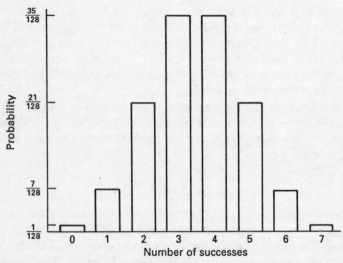

*Figure 5.1.* Binomial distribution: $n = 7$, $p = \frac{1}{2}$.

Figures 5.1 and 5.2 are graphical equivalents of Tables 5.2 and 5.3.

## The mean or expectation

It is useful to have summary figures that tell us where a distribution is *centred* and how much it *spreads*. A number of measures are available; we concentrate on one for *centrality* and two (closely related) for *spread*.

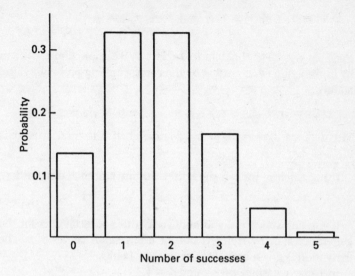

*Figure 5.2.* Binomial distribution: $n = 5$, $p = \frac{1}{3}$.

We measure *centrality* of a distribution by its *mean* or *expectation*, usually denoted by $E(X)$. If a random variable $X$ takes a fixed set of discrete values (e.g. as in the binomial distribution), the mean is obtained by multiplying each possible value by its probability, then adding these products.

*Example 5.6.* If we cast a true die, each of the outcomes 1, 2, 3, 4, 5, 6 has probability $\frac{1}{6}$. Thus, the mean, which is often written $E(X)$ (the 'E' for expectation), is given by

$$E(X) = 1 \times \tfrac{1}{6} + 2 \times \tfrac{1}{6} + 3 \times \tfrac{1}{6} + 4 \times \tfrac{1}{6} + 5 \times \tfrac{1}{6} + 6 \times \tfrac{1}{6}$$
$$= \tfrac{21}{6} = 3.5.$$

If, on the other hand, we had a biased die with probabilities given below (note these probabilities add to 1):

| $X$ | 1 | 2 | 3 | 4 | 5 | 6 |
|------|------|------|------|------|------|------|
| Prob. | $\frac{1}{21}$ | $\frac{6}{21}$ | $\frac{2}{21}$ | $\frac{5}{21}$ | $\frac{4}{21}$ | $\frac{3}{21}$ |

then

$$E(X) = 1 \times \tfrac{1}{21} + 2 \times \tfrac{6}{21} + 3 \times \tfrac{2}{21} + 4 \times \tfrac{5}{21}$$
$$+ 5 \times \tfrac{4}{21} + 6 \times \tfrac{3}{21} = \tfrac{77}{21} = \tfrac{11}{3} = 3.67.$$

*Example 5.7.* For the data in Table 5.2, we could find the mean by multiplying each $r$ value by the corresponding probability and adding:

$$E(X) = 0 \times \tfrac{1}{128} + 1 \times \tfrac{7}{128} + \cdots + 6 \times \tfrac{7}{128} + 7 \times \tfrac{1}{128}.$$

This tedious exercise is not recommended! The final answer is $E(X) = 3.5$.

Using algebra, we can show that, for any binomial distribution,

$$E(X) = np. \tag{5.8}$$

If you enjoy algebra, you may verify this yourself either for the general case or by trying it out for a few small values of $n$. The derivation is given in most statistics text books.

In Table 4.3, since $n = 7$ and $p = \tfrac{1}{2}$,

$$E(X) = 7 \times 0.5 = 3.5.$$

For the example in Table 5.3 ($n = 6$, $p = \tfrac{1}{3}$), $E(X) = 6 \times \tfrac{1}{3} = 2.0$. Note that the mean is *not* a typical value of $X$ in the sense that it need be attainable. Families do not have 3.5 children!

*Example 5.8.* For the wager considered in Example 4.10, de Méré's *gain* on a unit stake is a random variable $X$ which takes the value 1 if he wins or $-1$ if he loses (regarding a loss as a negative gain). The probability that he wins is 0.5177; that he loses is 0.4823. Thus, his expected gain per play is

$$E(X) = (1 \times 0.5177) + (-1 \times 0.4823) = 0.0354.$$

In one hundred plays the expected gain is a hundred times that at a single play (i.e. £3.54), as Pascal predicted (see p. 56).

Similar arguments for the wager in Example 4.11 show de Méré has a negative expected gain (i.e. a loss) of £1.72 per 100 games.

*Example 5.9.* In a charity raffle you are invited to pay an entry fee of £5 which allows you to cast a die six times. If you throw six

6's you win a £10 000 car (and your £5 entry fee is refunded!). What is your expected gain?

Clearly, the probability you gain £10 000 (in the form of a car) is $(\frac{1}{6})^6$ (by the multiplication rule). The probability you lose £5 (the opposite event) is $1 - (\frac{1}{6})^6$. The expected gain per play is therefore

$$[10\ 000 \times (\tfrac{1}{6})^6] + \{-5 \times [1 - (\tfrac{1}{6})^6]\} = -4.79.$$

In other words the expected *gain* to the charity (or loss to you) is £4.79 per play. It would of course be too bad for them if the first player won the car. However, that only has probability $(\frac{1}{6})^6 = 0.000\ 021\ 4$. More elaborate rules give the expected number of attempts before the prize is won. We omit details, but suffice to say most charities do pretty well out of this type of lottery, especially as the car is usually donated.

## Why Paul was robbed

In Example 3.11 we looked at a game in which Paul paid Peter £$2^k$ if the first head occurred at the $k$th toss of a coin. It is easy to see that the probability of the first head at the first toss is $\frac{1}{2}$, at the second toss $(\frac{1}{2})^2$ and, more generally, at the $k$th toss $(\frac{1}{2})^k$, since we have a sequence of $k - 1$ tails, then one head. The expected pay out by Paul at any play of the game is obtained by multiplying each probability by the corresponding payout and adding. Thus, the expected payout is

$$2 \times \tfrac{1}{2} + 2^2 \times \tfrac{1}{2^2} + \cdots + 2^k \times \tfrac{1}{2^k} + \cdots,$$

where $k$ may take any integral value. Note that each of these terms has the value 1, so the sum becomes

$$1 + 1 + \cdots + 1 + \cdots.$$

This sum has no limit, since we may need an indefinitely large $k$ before getting a head. This means Paul's expected payment to Peter at each play is infinite. In return he receives but £8 per play. Inevitably, in many games Paul will lose all his money; if he has plenty of capital, this may take a long time. We saw in Example 3.11 that even a few games may lead him into trouble. The moral: if you are a gambler, don't let intuition be your master.

## Variance and standard deviation

We measure spread, or scatter about the mean, by *variance* or *standard deviation*. The variance, in essence, is the expectation, or mean, of the squares of deviations of observations from the mean itself. It is usually denoted by var($X$). In other words var($X$) is the average squared deviation from the mean. It sounds a mouthful, but an example explains it.

*Example 5.10.* For the fair die considered in Example 5.6, the mean is 3.5 and the deviations of scores from 3.5 are $1 - 3.5 = -2.5$, $2 - 3.5 = -1.5$, and continuing this way $-0.5$, $+0.5$, $+1.5$, $+2.5$. We square these and multiply each square by $\frac{1}{6}$ (the probability associated with each score) and add all the products to get the variance. It is 2.92.

The square root of var($X$) is an index of average deviations from the mean called the *standard deviation*. For the true die, it is approximately 1.7.

For some specific distributions, var($X$) has a simple form; we shall content ourselves with quoting appropriate formulae when we need them. For the binomial distribution with $n$ observations and probability of success $p$, the variance of $X$, the number of successes, is

$$\text{var}(X) = npq.$$

Thus, for Table 5.2, where $n = 7$ and $p = q = \frac{1}{2}$, we have

$$\text{var}(X) = 7 \times \tfrac{1}{2} \times \tfrac{1}{2} = 1.75.$$

For the distribution in Table 5.3, where $n = 5$, $p = \frac{1}{3}$ and $q = \frac{2}{3}$, we find

$$\text{var}(X) = 5 \times \tfrac{1}{3} \times \tfrac{2}{3} = \tfrac{10}{9}.$$

In general, the larger var($X$) (or the standard deviation), the greater the spread of the distribution.

## Some other distributions

The binomial distribution is an example of a *discrete* distribution,

*Table 5.4.* Total scores for each of thirty-six
outcomes with two dice

| | Score die 1 | | | | | |
|---|---|---|---|---|---|---|
| Score die 2 | 1 | 2 | 3 | 4 | 5 | 6 |
| 1 | 2 | 3 | 4 | 5 | 6 | 7 |
| 2 | 3 | 4 | 5 | 6 | 7 | 8 |
| 3 | 4 | 5 | 6 | 7 | 8 | 9 |
| 4 | 5 | 6 | 7 | 8 | 9 | 10 |
| 5 | 6 | 7 | 8 | 9 | 10 | 11 |
| 6 | 7 | 8 | 9 | 10 | 11 | 12 |

so-called because $X$ takes only a finite set of specified values. Distributions of this type are often associated with counts or scores and typically, though not essentially, take integral values. Measurements such as weights, times to failure, etc. are usually *continuous* random variables, it being possible for them to take any value within a certain range (providing we measure with sufficient accuracy). We meet some continuous random variables in Chapter 6. Here are two more examples of discrete distributions.

*Example 5.11.* Two dice are cast. What is the distribution of the sum of the scores?

Any of the six possible scores (1 to 6) with the first die may be combined with any of the six for the second die, giving thirty-six equally likely configurations with total scores between 2 (corresponding to a double 1) and 12 (corresponding to a double 6). The totals are conveniently represented in Table 5.4, the entries in the body of the table being the sum of the scores shown at the top of that row (die 1) and at the left of that column (die 2). The cases are represented pictorially in Figure 5.3.

In the table results giving equal scores line up diagonally, in parallel, from top right to bottom left. We see, for example, that four out of the thirty-six equally likely outcomes give a score of 5, another three a score of 10, and so on. If $X$ represents total score, we thus have $\Pr(X = 5) = \frac{4}{36} = \frac{1}{9}$ and $\Pr(X = 10) = \frac{3}{36} = \frac{1}{12}$.

*Table 5.5.* Distribution of total scores when a die is cast

| Total score: | 2 | 3 | 4 | 5 | 6 | 7 | 8 | 9 | 10 | 11 | 12 |
|---|---|---|---|---|---|---|---|---|---|---|---|
| Probability: | $\frac{1}{36}$ | $\frac{2}{36}$ | $\frac{3}{36}$ | $\frac{4}{36}$ | $\frac{5}{36}$ | $\frac{6}{36}$ | $\frac{5}{36}$ | $\frac{4}{36}$ | $\frac{3}{36}$ | $\frac{2}{36}$ | $\frac{1}{36}$ |

Considering each outcome in this way, we get the distribution of total scores given in Table 5.5.

We may calculate the mean and variance of this distribution directly. Multiplying each score by its probability and adding gives the mean value 7. For any *symmetric* distribution, the mean is equal to the middle value – one of the reasons we speak of the mean as an *indicator of centrality*. This makes it easy to write down immediately the mean for any symmetric distribution.

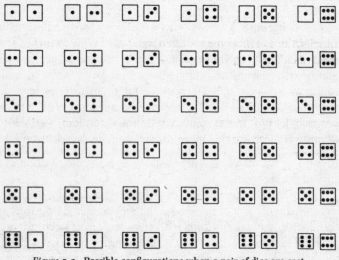

*Figure 5.3.* Possible configurations when a pair of dice are cast.

Note that it is twice the mean score of 3.5 on one die; there is a general result that, if we add two random variables like the scores on each die, the mean of the sum is equal to the sum of the means.

## Calculating a variance

The bald statement that the variance is the mean of the squares of deviations from the mean gives one way to calculate variance, but there is a better method. First, we define the mean value of the square of a random variable $X$ in a similar way to the mean value of $X$ itself, except that in the sum of products we replace each value of $X$ by its square. We write this $E(X^2)$ and for the score on a pair of dice example, using Table 5.5,

$$E(X^2) = (2^2 \times 1 + 3^2 \times 2 + 4^2 \times 3 + \cdots + 12^2 \times 1)/36$$
$$= \tfrac{329}{6}.$$

To obtain $\text{var}(X)$, we use the result

$$\text{var}(X) = E(X^2) - [E(X)]^2,$$

where $[E(X)]^2$ is the square of the mean value of $X$. Since $E(X) = 7$, we get $\text{var}(X) = \tfrac{329}{6} - 49 = \tfrac{35}{6}$.

## An empirically based distribution

We may build up distributions using the *relative frequency* concept of probability. We study, often over a long period, the relative frequencies of all possible outcomes of an experiment. Here is a simple example. A machine has an oil reservoir holding 200 gallons. Each day it uses between ten and twenty gallons of oil. The amount it uses on any day is not predictable in advance but depends on demands put upon it. The reservoir oil level is checked each morning and it is refilled if there is twenty gallons or less remaining. If the maximum consumption of twenty gallons were used every day, refilling would be needed on day ten, as 180 gallons would have been used during the previous nine days, leaving only twenty in the reservoir. If, on the other hand, the minimum of ten gallons were used each day, filling would not be required until day nineteen. Records taken over a long period (or with large numbers of identical machines) will show the relative frequencies of days to refilling. These are used to form an *empirical* distribution. The word *empirical* means experimentally observed.

*Table 5.6.* Relative frequencies of run times between oil refills

| Days, X: | 10 | 11 | 12 | 13 | 14 | 15 | 16 | 17 | 18 | 19 |
|---|---|---|---|---|---|---|---|---|---|---|
| Rel. freq.: | 0.04 | 0.07 | 0.08 | 0.09 | 0.12 | 0.15 | 0.19 | 0.12 | 0.08 | 0.06 |

*Example 5.12.*    Records of the numbers of days a machine ran before the oil level dropped to twenty gallons in the circumstances described above, based on 5000 refills, gave the relative frequencies of X, the number of days to refilling, in Table 5.6. Note that the sum of the relative frequencies, taken as probabilities for the empirical distribution, is 1. We calculate the mean, variance and standard deviation in ways already described. Calculations (it is worth doing these as a check) give the mean as 14.89, the variance as 5.6579 and the standard deviation as 2.3786. Thus, the average number of days between refills is about fifteen with a standard deviation (our index of average variation) of just over two days.

## Summary

1. The number of permutations (order matters) of $r$ items from $n$ is $^nP_r = n \times (n - 1) \times (n - 2) \times \cdots \times (n - r + 1)$.

2. The number of combinations (order ignored) of $r$ items from $n$ is $^nC_r = {}^nP_r/r!$.

3. Bernoulli trials have two outcomes, success (probability $p$) and failure (probability $q = 1 - p$). The number of successes $r$ in $n$ independent Bernoulli trials is a random variable $X$ with $\Pr(X = r) = {}^nC_r p^r q^{n-r}$ $(r = 0, 1, 2, \ldots, n)$. The random variable $X$ has a binomial distribution.

4. The mean or expectation $E(X)$ measures centrality. To form $E(X)$ for a discrete random variable multiply each possible value of $X$ by its probability and add all products. For a binomial distribution, $E(X) = np$. For a symmetric distribution, $E(X)$ is the middle value.

5. The variance and standard deviation measure spread. The variance is the expectation of the squares of deviations from the

mean, best calculated as $\text{var}(X) = E(X^2) - [E(X)]^2$. For a binomial distribution, $\text{var}(X) = npq$. The positive square root of $\text{var}(X)$ is the standard deviation.

# CHAPTER 6

# PROBABILITY IN ACTION

## Small changes

Risk situations often involve unpleasant events that have a small probability. About one heavy smoker in 600 dies of lung cancer. Annual fatalities in relatively dangerous occupations seldom exceed a few hundred per million workers. In many countries infant mortality (deaths at age under one year) is between six and twenty per 1000 live births.

Instinct to survive alerts us to changes in incidence of unpleasant or dangerous events. Increases in morbidity or mortality from industrial pollution, or side effects of a drug, are often associated with a small change in an already small probability.

Such changes are hard to detect. Records may show that *approximately* one person in 125 is affected each year by a virus disease and that it seems to be simply a matter of chance (perhaps associated with environmental, genetic or physiological factors) whether or not an individual succumbs. For ailments that are not infectious or contagious, with an incidence rate of one in 125, if we select a sample of a hundred people at random (p. 107), then $X$, the number in the sample with the disease, has a *binomial distribution* (p. 78) with $n = 100$ and $p = \frac{1}{125} = 0.008$. The *expected number* $E(X) = np = 100 \times 0.008 = 0.8$. Substituting $n = 100$, $p = 0.008$ and $q = 1 - p = 0.992$ in expression (5.6) on p. 79, we get the probabilities that $X$ takes values 0, 1, 2, 3, ... For example,

$\Pr(X = 0) = {}^{100}C_0(0.008)^0 \times (0.992)^{100} = (0.992)^{100} = 0.448$. A calculator with a $y^x$ key helps.

Table 6.1. Probabilities of r diseased when $n = 100$ and $p = 0.008$

| Number, $r$: | 0 | 1 | 2 | 3 | 4 | 5 |
|---|---|---|---|---|---|---|
| $Pr(X = r)$: | 0.448 | 0.361 | 0.144 | 0.038 | 0.008 | 0.001 |

Thinking of probability as a limiting relative frequency (p. 57), we would find that in the long run a proportion 0.448 (i.e. 44.8%) of all samples of a hundred would contain no cases of the disease.

Table 6.1 gives the probabilities for various numbers diseased in a sample of a hundred. The sum of the probabilities in Table 6.1 is unity. Even allowing for rounding off, it is clearly unlikely there will be more than five cases; indeed, the probability of six or more cases is only 0.000164. We would certainly not expect fifty cases in a sample of a hundred (50%) if the incidence rate were 0.008 (less than 1%) in the population as a whole.

## Ninety-five per cent range

Table 6.1 shows that zero, one or two are the most likely numbers, and the associated total probability is $0.448 + 0.361 + 0.144 = 0.953$. This means that if we took a lot of samples of a hundred, then in more than 95% of them we would find between zero and two diseased. We call this set of *most likely* outcomes with associated probability *at least 0.95* a *95% range*. The idea is widely used, but the term 95% range is not universal.

## A higher incidence

If the incidence rate jumps from one in 125 to one in a hundred, the number of cases $X$ in samples of a hundred has a different binomial distribution; we still have $n = 100$, but now $p = 0.01$. Table 6.2 gives probabilities for zero to seven cases. Adding probabilities over zero to three cases, we get a total of 0.981. These outcomes form the 95% range (no shorter range has associated probability 0.95 or more); it overlaps the range zero to two when $p = 0.008$. Clearly, no matter whether $p = 0.008$ or $p = 0.01$,

*Table 6.2.* Probabilities of r diseased when $n = 100$ and $p = 0.01$

| Number, r: | 0 | 1 | 2 | 3 | 4 | 5 | 6 | 7 |
|---|---|---|---|---|---|---|---|---|
| $Pr(X = r)$: | 0.366 | 0.370 | 0.184 | 0.061 | 0.015 | 0.003 | 0.001 | 0.000 |

*Table 6.3.* Probabilities of r diseased when $n = 100$ and $p = 0.02$

| Number, r: | 0 | 1 | 2 | 3 | 4 | 5 | 6 | 7 | 8 |
|---|---|---|---|---|---|---|---|---|---|
| $Pr(X = r)$: | 0.133 | 0.271 | 0.273 | 0.182 | 0.090 | 0.035 | 0.011 | 0.003 | 0.001 |

we will often get between zero and two cases in a sample of a hundred. Even if the incidence rate now doubles to $p = 0.02$, with samples of a hundred ($n = 100$) we get the probabilities in Table 6.3. The 95% range is now zero to four, and we would still expect to get only one case per hundred in over one-quarter (in fact, 27.3%) of a large number of samples, so it is clear that with just one sample of a hundred it is often impossible to distinguish between incidence rates as diverse as one in 125 and one in fifty.

Of course, if we found twelve cases in a hundred, we would take little convincing that the incidence rate was higher, perhaps about one in eight or one in ten; but dramatic changes in rates are usually no problem, being easily spotted by a sharp rise in cases seen by G.P.s or admitted to hospital. Fortunately, larger samples help us spot small changes.

### A yardstick

The 95% range provides a yardstick for choosing between different $p$ values. These ranges depend on the sample size $n$ as well as on $p$.

For a fixed value of $n$, we may calculate the 95% range for any $p$ we are interested in. If then, in a sample of $n$, the number of cases we observe lies in the 95% range for a specified $p$, we say that value of $p$ is *acceptable* as a *possible* incidence rate.

To choose between two possible values for $p$, we ideally want non-overlapping 95% ranges to decide which is appropriate. We have seen we do not achieve this with $n = 100$ if $p = 0.01$ or $p = 0.02$.

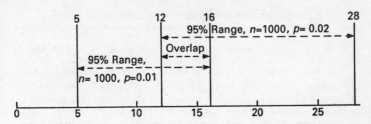

*Figure 6.1.* 95% ranges showing overlap for two *p* values.

What if $n = 1000$? Tedious calculations of binomial probabilities (replaced in practice by good approximations) show that, when $n = 1000$ and $p = 0.01$, the 95% range is five to sixteen, inclusive; if $p = 0.02$ it is twelve to twenty-eight. These overlapping ranges are shown in Figure 6.1.

For any particular sample, we observe only one value. If it were seven, say, this would suggest $p = 0.01$; had it been twenty-two, this would point to $p = 0.02$. If it were in the overlap part of the ranges (i.e. between twelve and sixteen, inclusive), then either $p$ value is acceptable on our 95% range criterion, and we say there is insufficient evidence to choose between $p = 0.01$ and $p = 0.02$.

## Larger samples

Still larger samples may resolve the issue. Would a sample of 10 000 suffice? Calculations (easy on a calculator using approximations to binomial probabilities and not described here, but familiar to many statistics students) give 95% ranges of eighty to 120 when $p = 0.01$ and 172 to 228 when $p = 0.02$. These do not overlap; if we get a number in one of these ranges and *these are the only possible values of p*, we would be virtually certain which were the true one.

Table 6.4 gives 95% ranges for four $p$ values, corresponding to incidence rates (a) one in 125, (b) one in a hundred, (c) one in eighty and (d) one in fifty, for samples of various sizes.

Range overlap is almost complete when $n = 100$; there is some

Table 6.4. 95% ranges for samples from specified binomial distributions

| | $n$ | | | | |
|---|---|---|---|---|---|
| | 100 | 1000 | 10 000 | 100 000 | 1 000 000 |
| (a) $p = 0.008$ | 0–2 | 2–14 | 62–98 | 744–856 | 7821–8179 |
| (b) $p = 0.01$ | 0–3 | 5–16 | 80–120 | 938–1062 | 9804–10196 |
| (c) $p = 0.0125$ | 0–3 | 7–19 | 103–147 | 1181–1319 | 12281–12719 |
| (d) $p = 0.02$ | 0–4 | 12–28 | 172–228 | 1913–2087 | 19723–20277 |

overlap between all four when $n = 1000$; still some between (a) and (b) and between (b) and (c) when $n = 10\ 000$, but (d) is now clear. All four are distinguishable when $n = 100\ 000$ or more.

Use of a 95% range is customary and conventional. Alternatives sometimes used are 90%, 99% or occasionally 99.9% to cover the most likely outcomes with total associated probabilities 0.90, 0.99 and 0.999, respectively; the higher the probability, the wider the range.

### Other incidence rates

How do we take account of other possible incidence rates? We can, for a given sample size $n$, calculate a 95% range for *any* incidence rate. Then, if, say, $n = 100\ 000$ and we observe 1300 cases, tedious calculations (we omit details because approximations are used in practice) show 1300 cases lie in the 95% range for incidence rates between one in 72.85 and one in 81.2. This important idea is given a special name. We say that, having observed 1300 cases in 100 000, the band one in 72.85 to one in 81.2 is a 95% *confidence interval* for the true incidence rate.

If we took another sample of 100 000, even if the unknown but true incidence rate remained the same, we would be unlikely to get again exactly 1300 cases; we might get 1284 or 1338 or 1311. For each of these, we would get a slightly different 95% *confidence interval*. The important property of these intervals, though we do not prove it, is that in the long run ninety-five out of every hundred of them will include the unknown *true* incidence rate.

The confidence interval provides what is called an *interval estimate* of the true incidence rate. The interval gets shorter with a larger sample, because having more data improves our estimate. In practice we obtain confidence intervals by less tedious methods, but these require further theory.

To sum up:

(i) For infrequent events, we need a large sample to pick up small changes in rate.
(ii) Even if we assert there is a change, there is still uncertainty about how big it is; a confidence interval gives us a likely range for the true value.

In Chapter 7 we describe famous experiments that required samples of about 1 000 000 to obtain important and useful results; things there were not as simple as in the examples above. Indeed, they seldom are. For infectious and contagious diseases, the mathematical model is more complicated than that for the binomial distribution. Some diseases show inexplicable variations in incidence; for leukaemia in children, we find local pockets of higher than usual incidence for at present unknown reasons. Some pockets are near nuclear processing plants; others are not.

## Accuracy of information

No statistical analysis can overcome serious data deficiencies. In studies of morbidity and mortality it is notoriously difficult to obtain *relevant and accurate* information. This is especially true of data from medical records or death certificates. Doctors make wrong diagnoses and state causes of death in vague terms. A death certificate may, quite correctly, indicate cardiac arrest as cause of death, yet give no indication that this might be a secondary consequence of a disease we are interested in.

Another problem for making inferences from samples is that the sample may not be selected in a way that ensures freedom from bias (see p. 47). Incidence rates for a disease obtained from school records, generally practitioners' case notes or hospital

Table 6.5. Numbers of city-dwellers responding to two drugs

|  | Standard | New | Totals |
|---|---|---|---|
| No effect: | 500 | 1050 | 1550 |
| Cure: | 100 | 350 | 450 |
| Total: | 600 | 1400 | 2000 |

admissions, clearly may not, and usually do not, reflect rates in the population at large.

### Interpretation

Even accurate data may show differences that need care in interpretation. For example, records based on large samples may show the death rate under one anaesthetic is three per 100 000 administrations; eight per 100 000 for another. This may be confirmed (using 95% ranges) as a real difference. *All other things being equal*, this implies a higher risk with the second anaesthetic. But are all other things equal? Doctors do not make random choices of anaesthetics. The second anaesthetic may be used mainly for patients with a cardiac abnormality, or for inherently dangerous operations. A higher death rate under anaesthetic is to be expected for such cases. The apparently more dangerous anaesthetic may be used in these cases because experience has shown it is the safest under those circumstances.

### Other inferences

The 95% *range* is a useful concept to quantify uncertainty in choosing between hypotheses (e.g. two different incidence rates) or in forming a *confidence interval*. To elaborate on this theme would make this book a statistics text, but we look at one more example, using a rather different approach and the data for *city patients only* from Table 3.4. These are reproduced in Table 6.5 with an added column giving totals for *no effect* and for *cure* for all patients (i.e. those receiving the standard and those receiving the new drug).

*Table 6.6.* Expected cures and failures if drugs
equally effective

|  | Standard | New |
|---|---|---|
| No effect: | 465 | 1085 |
| Cure: | 135 | 315 |

We ask: 'Is one of the drugs (standard or new) really more effective, or could any difference in observed cure rate be attributed to chance?'

Put another way, are we likely to get as great or a greater observed difference between proportions cured by each drug *if they are really equally effective?* If the probability of doing so is small, it suggests a real difference. This is a problem of hypothesis testing. We set up as our 'cock-shy' the hypothesis, often called a *null hypothesis*, that there is no difference.

We proceed this way. Among all 2000 patients there are (last column of Table 6.5) 450 cures and 1550 'no effects'. If there is really no difference between standard and new drug, we would expect roughly the same proportion of cures among the 600 receiving the standard drug and among the 1400 receiving the new. We can work out the expected numbers in each case. The argument is simple, but needs following carefully. A pencil and paper may help you do this. Since 1550 out of 2000 show no effect, among 600 (the number given the standard drug) we would *expect* $600 \times \frac{1550}{2000} = 465$ to show no effect. This means that among 600 we could expect $600 - 465 = 135$ cures. Also, since we expect, under the hypothesis of *no difference*, 465 out of 1550 'no effects' to occur with the standard drug, we must expect the remainder of that 1550 (i.e. $1550 - 465 = 1085$) to occur with the new drug. Again, since we expect 1085 'no effects' among the 1400 who receive the new drug, we expect $1400 - 1085 = 315$ cures. These expected numbers are given in Table 6.6. You will see that the standard drug cures less than expected (a hundred compared with an expected 135), the new drug more. Large discrepancies between corresponding entries in Tables 6.5 and 6.6 cast doubt on the null hypothesis of equal effectiveness. The measure of discrepancy used to test for a real effect is to calculate for each of the four 'cells', or entries, corresponding to standard/no effect, standard/cure, new/no effect and new/cure the quantity

$$\frac{(\text{observed number} - \text{expected number})^2}{\text{expected number}}.$$

For the cell 'standard/no effect', the value is

$$\frac{(500 - 465)^2}{465} \approx 2.63.$$

For the remaining three cells, the corresponding values (you should check) are new/no effect 1.13, standard/cure 9.07 and new/cure 3.89. We add these (i.e. $2.63 + 1.13 + 9.07 + 3.89$) to obtain 16.72. Clearly, the higher this total, the stronger the evidence the drugs are of differing efficacy, for we expect each difference *observed* − *expected* to be small if the treatments are equally effective. Tables like those in Neave (1981: p. 21) give the probability of getting certain observed, or higher, values of this total when the hypothesis *no effective difference* is true. If this probability is small (the odds against it are high), we reject the hypothesis of no effective difference (the *null hypothesis* in statistical jargon). This does not *prove* there is a difference; it simply indicates the odds against there being no effect are high. Indeed, tables show there is less than one chance in 1000 of getting a total greater than 10.83 when the null hypothesis holds. Our total 16.72 is well above this, so it is *almost* 100% certain that the new treatment really gives better results.

If the probability of getting as high a value as we observe for the test statistic is $\frac{1}{20}$ or less when the null hypothesis is true, we say we reject the hypothesis at the 5% significance level, or that the difference (between treatment effects) is *significant at the 5% level* (a somewhat overused piece of jargon). If the probability is $\frac{1}{100}$ or less we say the result is significant at the 1% level, or *highly significant*.

If we apply the same test to the country cases only in Table 3.4, evidence again supports the new treatment.

Had the standard treatment been superior to the new, we would also have got a high value for óur test criterion. The direction of departures from average proportions (to a higher proportion of cures) indicates the new treatment is more effective in our example. On p. 48 we noted that combining results for city and country patients gave an indication the new drug was *less* effective. We explained why this is a spurious finding on p. 49; the proper tests given here show the new drug is more effective in both city and country.

### Uncertainty and proof

There is a fundamental difference between uncertain situations and completely deterministic ones. In the latter, hypotheses are either true or false. The angles at the base of an isosceles triangle are equal, full stop. In the presence of uncertainty we can only make statements with a high degree of confidence they are correct (e.g. a 95 or 99% confidence interval) or a small probability they may be wrong (e.g. when rejecting at a 5 or 1% significance level). We do not prove or disprove hypotheses in the *absolute* sense, like we do in deterministic situations.

The above is a sketchy treatment of estimation and hypothesis testing, but suffices for the examples in this book. The practising statistician takes these matters further and treats them more rigorously.

### An example from parapsychology

Do you believe in clairvoyance or thought transfer? In 1968 a parapsychologist, J. Barry, experimented to see whether individuals could retard the growth of fungi by 'willing' them to grow more slowly.

It was a well-conducted experiment. Ten individuals took part. Each was put in a room with a different set of ten identical fungi cultures. Each person's set of ten cultures was randomly divided into two groups of five – one labelled *experimental*; the other *control* – and each subject was instructed to concentrate his thoughts on the experimental cultures and 'will' them to grow more slowly than the controls. No subject was allowed to touch any cultures.

If subjects had no powers of thought transfer, one would expect the experimental cultures to show less growth in about half the cases and more in about half. Measurements showed that, for nine of the ten subjects, the average growth of his experimental cultures was less than that of the controls. The appropriate hypothesis test (not described here) indicates this result is *significant*, so we reject the null hypothesis of *no power of thought transference*, implying that the subjects had a power to retard growth by appropriate thought processes. Does the result surprise you?

### Significance and belief

Remembering that there is a one in twenty chance of getting a significant result even if the null hypothesis (here that there is no thought transference power to retard growth) is true, the cynic might say this experiment was such a one in twenty case. The scientist's best tactic to dispel such a doubt is to repeat the experiment a number of times. If he gets a significant result in all, or a great many, of these repetitions, the case for rejecting the null hypothesis becomes much stronger. The situation has similarities to the idea of duplicating or triplicating fail-safe devices discussed in Example 4.9.

There are recorded cases where statistical analyses of experimental results have shown a new treatment to give significantly better results than a standard one, but later use of that new treatment on a large scale has shown it is no better. Sometimes this is because we have a situation where a one in twenty outside chance comes up in the initial experiment; such is the nature of uncertainty.

A tendency to publish too many apparently significant results arises because editors of scientific journals are more inclined to accept papers reporting significant results than those that do not. Thus, if, say, fifty experimenters all test a new drug which is really ineffective we might expect one, two or even three of these to get a significant result (the one in twenty chance). These may be the only ones reported in scientific journals, giving a biased picture. Fortunately this tendency is largely self-correcting, because, once a result claiming significance is published, if there are others who did the same experiment and found nothing significant (in the statistical or even the clinical sense), they will now contradict the finding of a beneficial effect. Indeed, when major decisions are involved, the days of isolated small experiments are probably numbered.

## More about distributions

### The Poisson distribution

In Chapter 5 we met the binomial and some other discrete distributions; another that occurs widely is the *Poisson distribution*. It is associated with random incidents. For example, in manufacturing nylon thread, faults may occur *randomly*; this means they occur independently of one another and there is a constant *mean number of faults per unit length*, that is, this mean number does not change with time or with the total length produced. The independence condition implies that the occurrence of one or more faults in a unit length does not influence the number in another unit length. If a unit length is 100 m and in a long production run the mean number of faults is 2.0 per 100 m, then the probability of observing 0, 1, 2, 3, 4, 5, ... faults in any 100 m length is given by a Poisson distribution with mean 2.0. For a given mean, the probability of any number of faults is calculated by a mathematical formula and these probabilities are tabulated by Neave (1981: pp. 14–16). If the mean number of faults is 2.0 per 100 m, the probabilities of getting various numbers of faults in 100 m lengths are given in Table 6.7. Compare these probabilities with those in Table 6.3. They are very similar. Indeed, a property of the binomial distribution is that, if $n$ is large and $p$ small, the binomial distribution, for which $E(X) = np$, differs little from the Poisson distribution with the same mean or expectation. In each of Tables 6.3 and 6.7 the mean is 2. Our results in Tables 6.1 to 6.4 would not

*Table 6.7.* Poisson probabilities, numbers of faults when mean is 2.0

| No. of faults, $r$: | 0 | 1 | 2 | 3 | 4 | 5 | 6 | 7 |
|---|---|---|---|---|---|---|---|---|
| $\Pr(X = r)$: | 0.135 | 0.271 | 0.271 | 0.180 | 0.090 | 0.036 | 0.012 | 0.003 |

be altered in any essential way if we based all probabilities on the appropriate Poisson distributions; also the calculations are then easier and the results are available in published tables.

## Continuous distributions

Random variables representing measurements usually have a *continuous distribution of values.* Weights, heights, thicknesses, lengths, blood pressure, time intervals, etc. may take any values within certain limits (if we measure them with sufficient accuracy).

## The normal distribution

The best-known continuous distribution is the so-called *normal distribution.* It is appropriate when a characteristic – be it the height of adult Caucasian males, or the weight of sugar bags filled by a machine, or the thickness of mass-produced laminate sheets – varies between individual 'items' in such a way that, if we calculate the mean $m$ (p. 83) and standard deviation $s$ (p. 86) for a large batch of size $n$ using the convention that each observation has an associated probability $1/n$, then it has the following properties:

(i) There is symmetry about the mean: half the values are above and half below the mean, the spread being the same either side of the mean.

(ii) Values of the characteristic close to the mean are more probable than values remote from the mean. The characteristic will take values lying (a) within one standard deviation from the mean for about two-thirds of all items; (b) within two standard deviations from the mean for approximately 95% of

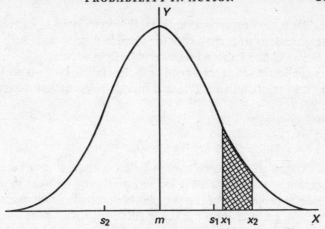

*Figure 6.2.* The normal distribution.

such items; (c) within 2.6 standard deviations from the mean for approximately 99% of such items. It follows from (i) that, for each of these ranges, roughly equal numbers will be above and below the mean.

(iii) Only about one item per thousand will have a value more than about 3.3 standard deviations above or below the mean.

These properties are not a formal mathematical definition of the normal distribution.

A useful graphical representation of this distribution is shown in Figure 6.2. The graph is called the probability density function of the distribution. The area under the curve is unity, corresponding to the total probability. The shaded area between any two $x$ values ($x_1$ and $x_2$) in Figure 6.2 gives the probability that $X$ takes a value in that range. The curve has its maximum at the mean $m$. The $X$ values $s_1$ and $s_2$ are each at distance $s$ (one standard deviation) from $m$.

Many properties of the normal distribution make it important in situations with uncertainty, both in hypothesis tests and for obtaining confidence intervals. Even when observations do not have a normal distribution, we may still use certain tests based on this distribution as approximations. It is even used to calculate approximations to probabilities for discrete distributions such as the binomial and Poisson under certain circumstances.

Its application requires a good deal of technical know-how. We meet it in relation to weights and measures legislation (Chapter 13, p. 235).

## Non-normal distributions

Many observations clearly do not follow a normal distribution. We might at 9 a.m. each day for a year at some weather station record the proportion of the sky that is cloudy. In many parts of the world, on most days the sky is either nearly all clear or else nearly all cloudy. This means that the mean or expected cloud cover may be half or three-quarters, but the numbers of days on which it is part cloudy are fewer than those when it is very cloudy or virtually cloudless. Extremes from the mean of *little or no cloud* or *much cloud* are then more common than values near the mean, unlike the situation for a normal distribution.

## Non-symmetry and long tails

Another important type of distribution covers non-symmetric situations. We meet these when we wait for a bus. The shortest time we have to wait is clearly zero minutes. If buses run frequently but are erratic because of traffic delays, we may find our mean (expected wait) is three minutes, yet, on occasions, because of heavy traffic or other delaying factors (e.g. collecting fares on a one-man operated bus, etc.), we may sometimes wait five, twelve, fifteen, seventeen, even twenty minutes.

The average life of a domestic cat is about eight years, but a few cats live to well over twenty years, and some even to thirty years. So *cat lifetime* has a non-symmetric distribution and, because some cats have extremely long lives, the distribution is sometimes referred to as 'long-tailed'. (This has nothing to do with the length of the cat's appendage!)

If we record the lifetime of vital machine components, the majority often last either their average life or just slightly more or less before they fail, but a few individual components may have a lifetime far in excess of average, giving a 'long tail' to the distribution of lifetimes.

A distribution called the *exponential distribution* is non-symmetric and long-tailed. If events occur randomly in time – this implies numbers occurring in a fixed time period follow the Poisson distribution (p. 103) – then the interval between successive events has an exponential distribution with mean equal to the average time between events. This distribution is important in studying the behaviour of certain queues.

### Samples

We have referred to *random* samples. These are samples from a large group of units (called a population) selected so that each unit in the population has an equal chance of being chosen. This protects against deliberate, or even unintentional, bias. In Example 3.10 we pointed out that, if choice of treatment were left to doctors, they might, for good reasons, always give one treatment to a particular type of patient (e.g. the more severe cases). This defeats the objective of testing a drug over a broad spectrum of cases.

There are sophisticated ways of improving upon random samples that are widely used in practice, but we need only the basic concept for this book.

In later chapters we look at situations where it is important that samples be random, or at least can be assumed free from bias of the kind likely with non-random selection. For a more detailed yet elementary discussion of samples and their properties, see *Understanding Data* (Sprent, 1988: Chaps. 14 and 15).

### Random numbers

To select random samples, also for studies of uncertainty using a technique called *simulation* that we describe in Chapter 10, we need sets of *random numbers*; these may be integers or fractions in a designated range. A set of twelve random integers in, say, the range one to 156 is so chosen that each integer in that range has the same probability of selection, that is, they are a random sample of the integers in the range. To get such a set, we could write down all the numbers, put them in a hat, shake them well and draw out our sample of twelve. Sometimes in sample selection we allow repeats; if so, after selecting each number we replace it and shake well before drawing the next. If we do not allow repetitions, no replacement is made after selection.

Drawing numbers from headgear may aptly be called 'old-hat'. Computers produce random numbers using *pseudo random-number generators*; 'pseudo' because they use a mathematical device rather than a physical generator to produce numbers that have all the properties of random ones.

## Building blocks

The key building block is the *random digit*, one of the digits 0, 1, 2, 3, 4, 5, 6, 7, 8, 9 chosen in such a way that each digit has the same chance of selection. Continued selection of digits this way builds up a sequence of random digits. They may be paired to form random numbers between 0 (written 00) and 99, or grouped in triplets to give random numbers between 0 (written 000) and 999. Most computers will also give selections of random integers in any required range, for example, ten random numbers (a random sample of ten) between 17 and 452. They also generate random fractions in a given interval (e.g. between 0 and 1).

Important properties of sequences of random digits are that

(i) in the long run each digit should occur approximately the same number of times;

(ii) the occurrence of a particular digit tells us nothing about the preceding or following digits.

Property (ii) means, for example, that, if we get a high digit, we cannot imply that the next will be low; or, if it is even, that the next will be odd, or anything of that sort. Indeed, even a run of high digits does not mean the next will be low. Each digit still has a probability of $\frac{1}{10}$ of appearing next. There are other properties concerning the frequency of runs of various lengths, of even or of odd digits, etc.

Here is a string of random digits generated by my computer (division into groups of five is simply for convenience in reading).

07206 61263 41881 26050 46753 34318 75391 92145
00912 27519 69376 14416 22488 51839 26010 21724
13735 15617

Random digits give a way of mimicking the happening of events, where each has some known probability of taking place. We meet examples in Chapter 10.

If a computer is not available to generate random numbers, we may use tables of random digits; see, for example, Neave (1981: p. 42).

### The role of statistics in uncertainty

In this chapter we have outlined uses of objective probability and statistics – what are now often regarded as classical statistical methods – to make inferences when there is uncertainty. These inferences usually take the form of estimating something (e.g. What is the incidence rate of a disease? What is the average blood pressure of individuals with some illness? What are the average number of underweight bags per 1000 bags of sugar?), or of testing a hypothesis (e.g. Has the incidence rate, the average blood pressure, the number of underweight bags changed because of some new circumstance – increased pollution, a new treatment, a new machine? Has compulsion to wear seat belts made the roads safer?)

Statistics is also the backbone for prediction and forecasting. Is the demand for electric power increasing? Is the annual rainfall showing a steadily decreasing trend? Is the ozone level in the atmosphere decreasing? One important forecasting technique is *time-series analysis*, where mathematical models are used to predict future behaviour on the basis of past observations (e.g. future demands of power on the basis of demand patterns over the previous twenty years).

Another important role for statistics is in modelling real-world processes in simulation studies. We shall meet elementary examples in Chapter 10, pp. 159–78. In particular, probabilistic models help us study effects of changes in a system. What is the effect of allowing longer intervals between appointments at a hospital out-patients clinic on patients' waiting time? What strain on facilities (equipment, doctor's work load, etc.) will occur if intervals are made shorter? Will we reduce the operating costs for a process by using new components that cost more, but give a longer average life?

## Summary

1. Small changes in low incidence rates can only be confirmed with large samples. The *95% range* is a useful concept for testing hypotheses about incidence rates and, more importantly, to obtain confidence intervals (interval estimates) for true rates. These concepts are widely used in other statistical inference problems.

2. The Poisson distribution is important in counts of random incidents and is closely related to the binomial distribution when $n$ is large and $p$ small.

3. Measurements of lengths, weights, etc. often follow continuous distributions, especially the normal distribution. The properties of the normal distribution give it a key role in statistical methods, but not all continuous data has (even approximately) this distribution.

4. Random samples (and more sophisticated types of samples) are used to avoid bias. Random digits (computer generated) are useful in forming such samples and also in simulation studies.

# PART II

TOURING AN UNCERTAIN WORLD

# CHAPTER 7

## POLIO: A $5 MILLION TEST

### The need for care

Major problems with built-in uncertainty may be national, international or global. In this chapter we tell the story of how one nation tested a new weapon to fight a crippling disease and give an account of classic experiments to evaluate a vaccine, showing the need for careful planning of a large trial. All this because the disease, although serious, affected only a small proportion of the population. We saw in Chapter 6, pp. 92–96, that large experiments are needed to detect real effects in such circumstances.

In this study, had care not been taken to demonstrate the efficacy of the vaccine beyond all reasonable doubt, there could well have been an emotional backlash against its use when, owing due to a later accident in its preparation, it caused an unfortunate outbreak of the very disease it was designed to prevent!

### Polio vaccine: a case study

#### The problem

Thanks to successful vaccines, poliomyelitis (polio) is now rare in developed countries, but in the early 1950s the number of diagnosed cases in the U.S.A. alone exceeded 30 000 per annum, a sixfold increase from the 1930s average of about 5000 per annum.

Despite this increase, the annual incidence rate in the early 1950s was still relatively low: about one case per 5000 people, roughly one-tenth the U.S. rate for deaths plus serious injuries in road accidents.

Three factors explained widespread concern about polio:

(i) Children in the five to nine age group were most vulnerable: polio caused some 6% of deaths in that age range.

(ii) A large proportion of those surviving were paralytic: some were life-long cripples; a few were able to breathe only with a respirator, the dreaded 'iron lung'.

(iii) An erratic epidemic nature bred apprehension: outbreaks were unpredictable in time or place. In one year Chicago might have hundreds of cases; New York virtually none. Nationwide, the number of cases varied substantially from year to year: about 30 000 in 1951, nearly twice that number in 1952 and a drop to about 37 000 in 1953.

This erratic uncertainty left many parents feeling helpless in their inability to protect their children; there was little they could do except resort to prayer.

### The Roosevelt factor

The quest for a vaccine was stimulated partly by the post-war rise in incidence, but also because thrice-President Franklin Delano Roosevelt was a polio victim, and his eminence ensured wide media publicity for its crippling nature. This is a *named life* factor of the type mentioned on p. 33.

To understand the problems in finding and testing a vaccine in the 1950s, we need to know more about the disease. Relevant points were:

(i) The number of *diagnosed* cases was relatively small, yet nearly everybody had at some time been infected by one of three strains of virus causing polio. Most people showed no clinical symptoms or only non-paralytic ones (e.g. slight fever or sundry aches) indistinguishable from symptoms of influenza and related-virus ailments. Many such victims never sought medical treatment; even had they done so, it is unlikely polio would have been diagnosed.

(ii) Once a person was infected, the system built up antibodies that usually gave lifelong immunity.

(iii) Incidence rates differed between races and communities, and within communities polio was more prevalent in sectors where hygiene and nutrition were *best*: this is why polio was dubbed a 'disease of civilization'.

In this last respect polio differs from most infectious diseases, where spread is associated with lack of hygiene. The contradiction is part explained by the way the body's immune system works. When hygiene is poor, the polio virus thrives and does indeed spread rapidly, so children are exposed to it as babies when they still have immunity from their mother's blood. This early exposure and universal presence of the virus builds a permanent immunity.

In more hygienic conditions the virus appears periodically and is widespread only during epidemics. In any community this may happen only every few years, the virus then attacking children who have lost immunity bequeathed with their mother's blood. For this reason polio as a disease was unknown in many African communities despite the presence of the virus. By contrast, symptomatic polio was widespread in North America, most of Europe, Australia and parts of Asia.

The mechanics of virus attacks, the unpredictability of epidemics or their severity, the difficulty of diagnosing 'mild' cases and the relatively low incidence rate, all led to problems in testing any vaccine appreciably less than 100% effective.

### Salk vaccine

Experiments with vaccination against polio started in the 1930s, when medical science was just beginning the serious study of viruses. These early experiments came to an abrupt halt when it became apparent that at least one vaccine under trial was capable of inducing paralytic polio.

After the war Jonas E. Salk of the University of Pittsburg identified three separate strains of virus causing polio. By 1952 he had developed a vaccine giving partial control of polio in monkeys. The primates were injected with suitably 'killed' viruses, and antibodies were formed that gave immunity to polio. As with many other 'killed' vaccines, this one was only effective if the virus was 'killed' in a special sense and three doses were administered at appropriate time intervals. The 'killing' agent was formaldehyde; the amount of 'killing' critical. 'Underkill' might allow the disease to develop in some who had been vaccinated; 'overkill', and the antibodies that provide protection would not be produced. By 1954 Salk and his co-worker Thomas Francis were ready to try the vaccine on humans.

They knew it would not be 100% effective, but hoped it would reduce the incidence by perhaps 50%, with a more substantial reduction in cases producing lasting paralysis.

### Small and variable numbers

Numbers of correctly diagnosed cases (mostly those showing characteristic symptoms such as paralysis) were small and variable even in the age group most at risk. So, as is implicit from our discussion in Chapter 6, pp. 92–96, a large number of children would have to be vaccinated to demonstrate any effect and to assess its likely magnitude.

The incidence rate in the U.S.A. each year in the early 1950s was usually *about* one case per 2000 in the five to nine age group, but remember that numbers diagnosed might halve or double from one year to the next, changes of the order of magnitude the investigators hoped to see in *any one year* by using the vaccine. Between localities, even in *one* year, differences were even greater: no cases in some areas; in others infection levels well above the national average.

Yearly variation ruled out testing the vaccine in one (or even several) years, and comparing incidence levels with those for pre-vaccine years. Likewise, using the vaccine in some cities or states but not others, then comparing results for *vaccinated* and *non-vaccinated* areas in the same year was a non-starter; one did not know in advance which areas would have epidemics, or the relative severity in different areas.

### Planned experiments

We explained in Chapter 3, p. 47, and Chapter 6, p. 107, that, if we want to compare two treatments (here vaccination compared to no action), it makes sense to split the subjects we study into two groups broadly alike in all ways *except* that one group receives vaccination and the other does not. In Example 3.10 we recommended *random* allocation to groups.

How big an experiment was needed to test the Salk vaccine? Would half a million children be enough? That would give two groups each of 250 000: one group to be vaccinated, the other not. With an infection rate of one in 2000, we would expect about 125 cases in 250 000. Annual and regional variations imply fewer cases in some years or communities and more in others, so we might get rather more or less than that expected number. Other factors causing variation are age, perhaps sex, social class and differing standards of diagnosis by doctors. These may influence numbers in a largely unpredictable way. Experience had shown that, even if the vaccine were completely ineffective, we might well get in one group as many as 140 cases and in the other perhaps only 110. Indeed, appropriate statistical theory shows that such a difference has more than one chance in twenty of arising by chance when there is no treatment effect – a result that is not significant at the 5% level (p. 100), so we could not reject the null hypothesis of no effect.

If we take certain precautions in allocating children to each group, then, using similar (but more sophisticated) ideas to those in Chapter 6, pp. 94–97, the statistician can tell us what difference we must have before we can reasonably conclude there is a beneficial effect of vaccination. Any smaller difference could be an artefact of chance; we would, at least, need to collect more evidence before claiming otherwise.

All this suggests half a million children may be too few. Indeed, two massive experiments involving nearly two million children were carried out at a cost of $5 million (a lot of money in the 1950s). This overcame many difficulties, while leaving administration of the vaccine practicable in terms of available resources.

## The experiments

We describe the experiments in the order they were planned; the second was developed as a result of reservations doctors and statisticians had about the first.

### Experiment 1

A number of schools were selected and agreed to take part. In each, children in grade II were offered vaccination. Because all vaccines carry a slight risk, for ethical and legal reasons vaccination was voluntary, requiring parental assent. All children (whether vaccinated or not) had their polio history for that year recorded. Was polio diagnosed? If so, how severe?

It might appear fair to test the effectiveness of the vaccine by comparing incidence rates for those inoculated and for those not inoculated, but there is a snag. The requirement of parental agreement resulted in a higher proportion of children with wealthier and better-educated parents being vaccinated. Better hygiene and nutrition implied a higher incidence of polio among such families. So this bias in selection was likely to *dilute* any evidence of a beneficial effect of vaccine, because a lower natural incidence rate could be expected among the unvaccinated. This is the sort of problem one may get when groups are not randomly selected.

To get a measure of bias, and to allow for it, at least in part, it was decided to follow up in all schools in the trial the polio histories of pupils in grades I and III, none of whom was offered vaccination. This left a difficulty of direct comparability, owing to the age dependence of polio incidence, but it was hoped that the effect of taking one grade with children generally younger and another with children generally older than those receiving the vaccine would be largely to cancel out the age effect.

There was one more worry. Difficulties of diagnosis might lead to different standards of reporting by each doctor and perhaps between districts; also doctors would soon learn that pupils in grade I and grade III had not had vaccine, while many of those in grade II had. Even subconsciously, some doctors might then consider more carefully their diagnoses for pupils in one group if they showed *any* suspicious symptoms because they felt ethically that they should take more care either in cases of vaccination (if they doubted its efficacy or even thought it harmful) or in its absence (if they felt such pupils more at risk). This meant each group may have been, on average, exposed to different standards of diagnosis. This is no criticism of the medical profession, but recognition of the 'human' element at work in the case of a difficult and important diagnosis.

### Experiment 2

Recognition of these difficulties stimulated a further trial. It required more organization, but eliminated the dangers of distortion due to

 (i) possible different average diagnostic standards for each of two groups;
 (ii) age differences;
(iii) the selective effect of volunteering.

It also allowed a direct investigation of any bias due to volunteers *only* being vaccinated.

Some three-quarters of a million pupils were approached. Of these, nearly 400 000 volunteered, with parental consent, to be vaccinated. They were randomly divided into two groups of roughly equal size. Remember what is meant by *randomly* (p. 107). Broadly speaking, in this case it meant using a mechanism effectively equivalent to tossing a coin to decide whether each child was allocated to group A or to group B. Further, and very importantly, neither any child nor any doctor knew to which group a child had been allocated; that was known only to the organizers. Those in group A were inoculated with the Salk vaccine. Those in group B were given a placebo, an inoculation *identical in appearance*, but consisting of a harmless and ineffective salt solution. Randomization should largely balance things like age differences, sex differences and differences in socio-economic group. It should also more or less even out any unsuspected factors that might influence susceptibility to the virus.

*Table 7.1.* Polio rates per 100 000

|  | All diagnosed | All polio | Paralytic | Non-paralytic |
|---|---|---|---|---|
| Salk: | 41 | 28 | 16 | 12 |
| Placebo: | 81 | 71 | 57 | 13 |
| Non-inoculated: | 54 | 46 | 36 | 11 |

### Double blind trials

The *double blind* procedure, as it is called, where neither doctor nor patient knows whether the active vaccine or a placebo has been administered, removes the possibility of diagnostic differences by individual doctors between pupils receiving the Salk vaccine and those receiving placebo, as they did not know which an individual had received. This 'double blind' approach is now widely used in clinical trials and is a legal requirement in many countries when testing new drugs using volunteer patients.

As inoculation was voluntary, there was a third group (nearly half the three-quarter million participants) who received neither vaccine nor placebo. Comparing this group with group B (placebo) would give a direct indication of bias due to volunteers from the higher socio-economic classes (and believed to be more at risk) predominating. For reasons given on p. 118, one would expect a higher incidence among those receiving the placebo than among those who did not volunteer.

### Interpreting results

If the two experiments gave broadly similar results and both favoured the vaccine, that would be strong evidence of its efficacy. If, however, there was a conflict of evidence, more weight could reasonably be given to indicators from the second experiment.

In all, 1 829 916 children were involved in the two experiments, 1 080 680 in Experiment 1 and 749 236 in Experiment 2.

Since there are fewer reservations about Experiment 2, we look first at the results for it. These are given in Table 7.1.

The figures given are numbers of cases *per 100 000 participating pupils*, that is, they are *incidence rates*. On the three lines, these are given separately for those receiving Salk vaccine, for those receiving the placebo and for those who did not receive any inoculation. Approximately 200 000 received Salk, a similar number the placebo and nearly 350 000 no injection (a few received a partial course of Salk or placebo, but not the full course of three injections).

The first column of the table gives number of cases diagnosed as polio per 100 000 children. Later laboratory tests showed not all diagnoses were correct: so the second column gives the incidence rate for correctly diagnosed cases; the third, the rate for those suffering paralysis; and the fourth, the rates for non-paralytic (i.e. milder) cases. Of course, we do not know how many undiagnosed cases there were, but presumably most of these showed only mild symptoms.

Looking at the lines corresponding to Salk and placebo, we see that Salk roughly halves the rate for all diagnoses compared with placebo, more than halves the rate for correct diagnoses and reduces it to less than one-third for paralytic cases. There is little difference in the rates for non-paralytic cases. The implication is clearly that Salk reduced the number of cases and also the severity for those who develop the disease. The effect of bias from volunteering is obtained by comparing the rates for the non-inoculated with those receiving the placebo. Since placebo and non-inoculated are for practical purposes equivalent treatments, this confirms our prediction that, because volunteers come from a predominantly higher risk class, their incidence rates without effective vaccination is higher. The exception is for non-paralytic cases, where the small difference in incidence rates could well be due to chance.

Table 7.2 is a corresponding table for Experiment 1.

There is reasonable agreement with broadly corresponding entries in Table 7.1. *Reasonable agreement* is admittedly vague. We use it to mean that they tell much the same story. Salk roughly

*Table 7.2.* Polio rates per 100 000

|  | All diagnosed | All polio | Paralytic | Non-paralytic |
|---|---|---|---|---|
| Salk (grade II): | 34 | 25 | 17 | 8 |
| Non-inoculated (grade I and III): | 61 | 54 | 46 | 8 |
| Non-inoculated: (grade II): | 53 | 44 | 35 | 9 |

halves the incidence rate and reduces the number of paralytic cases to about one-third. The bias due to using volunteers (comparing lines 2 and 3 in Table 7.2) again appears. Note, however, that the groupings in Tables 7.1 and 7.2 based on the different vaccination regimes are broadly, but not completely, comparable.

These two separately conducted, yet interrelated, experiments confirmed beyond all reasonable doubt (i.e. the results were statistically highly significant, p. 100) that Salk vaccine was effective, and led to programmes for mass immunization in many parts of the world.

### The follow up

This was not quite the end of the story. The experiments gave no hint of risk due to the vaccine, but there was a later setback when a batch of vaccine suffering from 'underkill' resulted in some children succumbing to polio as a result of vaccination. This is a typical example of a risk associated with a new, but generally beneficial, technological advance. There was great (and justifiable) public concern when this happened; concern that might have thrown back for years the fight against polio had the vaccine not been shown in carefully conducted experiments to be clearly beneficial. The vaccination programme might even have been abandoned. Improved manufacturing techniques quickly eliminated the difficulties. New methods using what are known as 'live' vaccines have been developed and have virtually eliminated polio where they are widely used. These vaccines contain attenuated strains of the virus that allow the antibodies to form without triggering the disease. One such vaccine was produced by Albert B. Sabin in 1960, bears his name and is administered orally, often on a sugar lump.

For simplicity we omitted one sidelight when discussing Experiment 2. Some doctors expressed disquiet about injecting a placebo, as this might be detrimental to any child who had already contracted polio but was not yet exhibiting symptoms. Medical experts rated any such risk as minimal if the injections were given when there was no current epidemic (summertime is the usual epidemic season). We mention this to emphasize that ethical aspects must not be forgotten when experimenting with humans (or animals!); it is important to minimize all known (or suspected) risks. With nearly every vaccine, there is a small element of risk, but, if the risk of serious damage from a vaccine is one in 1 000 000 and the risk of serious damage from the disease it controls is one in 10 000, it is generally regarded as justifiable to use an effective vaccine. Of course, the question of compensation for those seriously injured by a vaccine has important legal, ethical and social implications. We cannot be sure that a child who suffers serious effects from a vaccine would have been particularly susceptible to the disease.

A fuller account of the Salk vaccine tests is given by Meier (1972) and he includes more detailed tables of results. A critique of the medical aspects and implications was given by Salk's co-worker (Francis, 1955) immediately after the experiment was concluded and the results had been analysed. Brownlee (1955) looked at the technical statistical aspects.

## A review

The Salk story is a classic example of application of the science of experimental planning and statistical analysis based on objective probabilities to a real-world situation. The problem was basically one of determining small changes in incidence rates, something discussed for simple situations in Chapter 6. The complicating factors here were the remarkable changes in incidence rates (i) from place to place, (ii) from year to year and (iii) in different socio-economic groups. In addition, as in nearly all medical experiments, ethical requirements had to be taken into account. There were also diagnostic difficulties, especially for milder cases.

Fortunately, there was good background information. The vaccine under test had shown no adverse effect or serious side effects in preliminary trials. A lot was known about how incidence rates for polio varied with age, and sound statistical advice enabled possible criticisms of the first experiment to be met by a more sophisticated second experiment. Decisions about the size of experiments needed proved to be well founded.

The hiccup, at a later stage, when an insufficiently 'killed' vaccine was released did not reflect upon the experiment itself, but gave a salutory warning of the need for care in production of drugs and vaccines.

This is perhaps the greatest success story in this book for coping with uncertainty by using established statistical methods when there is adequate background information.

CHAPTER 8

# POISON IN THE AIR

### Atmospheric pollution

The Salk experiments epitomized one nation's success in establishing the value of a vaccine: the investment, $5 million; the spin-off,
worldwide. The solution – apart from that one incident with faulty
vaccine at a later date – was satisfactory, and there was little
cause for controversy. The main regret today is that the vaccine is
not freely available in some places where it is now most needed.

Polio was a matter of worldwide concern, but other international problems have a dimension that rules out one-nation
action for a solution. Their pervasiveness demands worldwide
collaboration and the final outcome may involve the balancing of
conflicting interests. Atmospheric pollution is one such problem,
a threat being tackled all too slowly for the following reasons:

(i) Despite increasing effort to collect relevant scientific information, there are still big gaps and a high uncertainty content.
(ii) Evidence is conflicting, some of it open to several interpretations.
(iii) The economic implications affect industries, governments
and international agencies. Coupled with this is a reluctance
to find money when others (some not yet born) may be the
prime beneficiaries.
(iv) Political will to take action is not strong because of this basically long-term aspect. One former U.K. Prime Minister will
probably be best remembered by posterity for the aptness of
his remark 'A week is a long time in politics'.

Testing the Salk vaccine needed statistically sound, but basically simple, though massive and expensive, experiments. Dealing with pollution problems is less straightforward: they are more complex, and we need first to disentangle the complexities of interacting chemical, meteorological, geological and biological factors, then to determine cost-effective means of reducing undesirable pollutants.

### Local pollution

Local pollution has been with us for centuries. An early hint of an international dimension came in 1691 when the diarist John Evelyn (1620–1706) wrote *Fumifigium*, a treatise on air pollution in London. He refers to complaints from the French of 'being infected with smoakes driven from our Maratime Coasts, which injured their Vines in Flower'.

Nearly 300 years later, 1987 was designated European Year of the Environment, springboard for a programme with a strong emphasis on studying, and tackling, pollution problems.

### Smog

London air pollution peaked in the smog of December 1952. In five days deaths attributable to the sulphurous, acid-laden, choking grey blanket that enveloped the city matched the numbers killed in road accidents for the whole of the U.K. for the complete year.

Then came the Clean Air Act, designed to ensure London would never suffer a repeat of that dreadful week in 1952, but still all is not well. Sooty smoke has gone, but less visible pollutants – ozone, acid rain, vehicle exhaust gases and associated products – are on the increase, eating into the fabric of buildings and putting stresses on the lungs and other vital organs of man and beast.

But pollution problems in London are a small part of those of the industrialized world. Just how things have developed since Evelyn wrote *Fumifigium* is told with startling clarity by Fred Pearce in *Acid Rain* (Pearce, 1987). This man-made curse has been publicized to the point where there is danger of blunting awareness by overkill, but, to twist a metaphor, acid rain is only the tip of the pollution iceberg.

## The scenario

Who are the culprits? In their primary form the pollutants are bulk-produced by

- public utilities (especially power stations)
- smelting, chemical and other industrial processes
- the internal combustion engine.

What are these pollutants? They are mainly

- sulphur dioxide (prime offenders thermal power stations)
- oxides of nitrogen; the so-called laughing gas is one of these! (power stations and transport)
- hydrocarbons (industry, transport)
- carbon monoxide (transport)
- heavy metals (transport).

In the atmosphere complex chemical reactions take place, giving rise to, among other things, the secondary products

- sulphuric acid
- nitric acid
- ozone.

Who and what suffers?

- forests (reduced growth, dead trees)
- waterways (lakes and rivers denuded of fish)
- agriculture (land acidification, reduction of crops)
- buildings (corrosion of stone and metals, flaking paint)
- people (health, irritation, inconvenience).

The tableau is illustrated diagrammatically in Figure 8.1.

### Some of the problems

Two secondary pollutants, water-soluble sulphuric and nitric acids, form the publicity-hogging *acid rain*, yet damage is probably

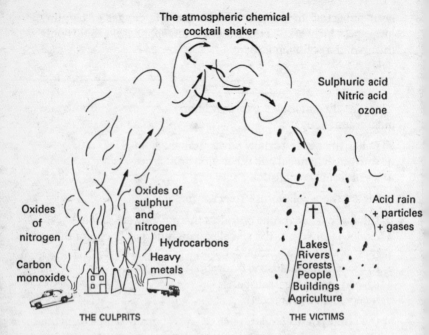

*Figure 8.1.* The saga of atmospheric pollution.

as great, perhaps even greater, from gaseous or particle poisons, these being about half the total deposition of pollutants. They are absorbed by plants (notably trees) and damage foliage; by soils to release toxic minerals; and in the fabric of buildings, corroding stone and metal. We inhale them to irritate our lungs and sensitive body membranes.

Contrary, then, to popular belief, *acid rain* is not synonymous with atmospheric pollution. It is indeed an unfortunate and ill-defined term, for rain always was acid. Pollutants have made our rain more acid; in extreme cases as acid as lemon juice. As well as acidifying lakes, rivers and soils, acid rain releases aluminium and other metals toxic to plants, at the same time washing out, spiriting away, beneficial soil components.

Pollution deposition varies from season to season, from year to year and from place to place. From rain, the damage is worst if soils are non-neutralizing or if the run-off from a watershed is so steep and fast that the acids find their way directly to rivers or lakes.

## Sources and types

Fossil fuels – coal, oil and natural gas – are our main sources of energy. Burning them produces those damaging pollutants, poisons widely dispersed in the atmosphere, windborne to points hundreds, even thousands, of kilometres from their source.

To distinguish liquid from gaseous or solid particle pollution, it is now customary to use the term *acid rain* for any wet depositions – rain, mist, fog, hail or snow.

In the U.K., Scotland has the worst acidification problems. The lakes of Galloway show evidence of acidification dating back to the industrial revolution, the process intensifying during the last few years, thanks to pollutants imported from England. The Scottish situation is not as bad as that in Scandinavia, but build-up effects are causing justifiable concern.

In the U.S.A. the most environmentally sensitive lakes and rivers are in the north-east. This is also their region of highest pollution output.

## Natural acidification

Not all acidification is due to pollutants. Natural biological processes form acids; also agricultural and forestry operations have profound effects locally on soil and water acidity. Thunderstorms, volcanic eruptions and wildfires are periodic and not very predictable natural causes of acidification that may affect large areas.

Offending industrialists have used natural acidification as a shield, a God-given excuse to shrug off the effects of their noxious outpourings. What they conveniently forget – if indeed they ever knew it – is that by its own counter-processes, nature has achieved for thousands of years a balance in the transport of natural nitrogen and sulphur compounds, forming the so-called *nitrogen* and *sulphur* cycles. Not perfect cycles: major, usually temporary, imbalances occur, but there is a long term global balance and only slow evolutionary changes. Four-fifths of the earth's atmosphere is nitrogen. This was not always so. It is likely that in days before life on earth most atmospheric nitrogen was in the form of ammonia. This reacted with carbon compounds to form substances known as amino acids and purines. The first primitive organisms may have evolved from these. A scientific account of such matters is given by Janet I. Sprent in *The Ecology of the Nitrogen Cycle* (Sprent, 1987).

### Man's interference

All the evidence points to man-made atmospheric pollution upsetting the natural cycles in a drastic, rather than evolutionary, manner.

The scale is not trivial. In one year man discharges into the atmosphere between thirty and fifty million tonnes of nitrogen oxides, and the figure is increasing. The natural output is uncertain (the processes are of great complexity): best estimates put it at only between one and three times this new man-made contribution. Industry discharges a hundred million tonnes of sulphur dioxide into the atmosphere each year; nature, half as much again.

Pollution is worldwide, from equator to poles, from sea level to the stratosphere. It is at its worst in highly industrialized areas, particularly in the northern hemisphere, where some 90% of the man-made contribution is released, nearly all in heavily industrialized belts in North America and Europe. Consequently, over these continents airborne man-made sulphur compounds heavily outweigh any from natural sources, including those somewhat unpredictable thunderstorms, volcanic eruptions and wildfires.

Pollutants are killing fish in the lakes and rivers of Europe and North America: triggering *Waldsterben* – a picturesque German term now used worldwide to describe forest decline; eroding the stone-work of historic cathedrals and bleaching their stained-glass windows; stripping protective paint from homes and offices; reducing crops; and affecting our health. Sometimes we know which of many components to blame for a particular evil; for others the evidence is conflicting. A question mark hangs over how far we can reverse damage already done. In all, a sorry saga of man's inability to cope with uncertainty, but there are lessons to be learnt, some already learnt, about how a threat with a multiplicity of causes and a multitude of different effects creeps up on us if we relax our vigilance and scientists, industrialists, politicians and their advisers fail to get their act together when it comes to decision making.

## Long-distance transport

Pollutants from, say, a coal-fired power station, may remain in the atmosphere for less than an hour or more than a week, be windborne across national frontiers, from continent to continent. Ironically, long-distance pollution has been aggravated by one step to reduce local pollution: taller stacks at power stations shoot polluting gases higher into the atmosphere. These stacks are memorials to short-sighted planning. They cut local pollution, but ignore the fundamental law that what goes up must come down – if not here, then somewhere else and perhaps in a different form – although we meet an alarming exception to this rule in the next chapter.

### Danger levels

Pollution levels and effects at points remote from the source of emission depend on

 (i) complex chemical changes that occur between emission and deposition;
 (ii) meteorological factors, especially temperature, air currents and humidity;
(iii) the nature of the earth's surface where the pollutants are deposited, whether as rain, dust particles or by adsorption of gases.

Details of the chemistry are ·not completely understood. The meteorological factors may be broadly predictable (prevailing wind from the west, higher humidity in the spring, etc.), but are subject to erratic variation. Atmospheric chemistry is a function of weather: sunshine and high temperatures encourage ozone formation; air currents mix chemicals from different sources in an active chemical cocktail. The geological aspect is important because some soils or rocks – especially limestones – act as neutralizing buffers to acids. There is a danger that, with increasing levels of pollutant, these buffers may eventually become ineffective. It will be hard to detect such a failure before it is too late.

Effects of acid rain on lakes and streams (and the fish in them) are well documented. Damage to forests, soils and buildings, and perhaps to human health, is attributable largely to secondary pollutants, generated in ways that are only slowly becoming clear. Nitrogen oxides and hydrocarbons react to produce ozone, a chemical toxic to plant life. We look at more fundamental problems of ozone balance in the atmosphere as a whole in the next chapter.

Added to clear dangers are inconvenient side effects; for example, sulphate particles in the atmosphere reduce visibility.

## The dangers of over-simplification

Uncertainty about precise effects and the paucity of information about movements of pollutants in the atmosphere leaves room for disagreement both about the true environmental risk and who should pay for controls. Evidence is often distorted, or interpreted differently, to support each particular case. Producers of pollutants play on the difficulty of pinning damage to any one source as an excuse to avoid, or at least postpone, costly reduction measures. Environmentalists, sometimes swayed more by understandable emotion than reason, are ever ready to seek scapegoats for any damage where pollution looks a likely, even a possible, culprit. They may conveniently forget the ravages of pests or disease that have nothing to do with the poisons in our atmosphere; ironically, some pollutants have a beneficial, if limited, effect in reducing pest populations.

Any cost–benefit analysis is fraught with difficulty. It is naïve to bandy round questions like 'Should ten million electricity users be prepared to pay 3% more for their electricity to protect sporting fishing on rivers?'

Reality is more complicated. While we continue to use fossil fuels, we will not eliminate all pollutants. So how effective will a costly partial reduction be? We might switch to low-sulphur coal at power stations or install expensive filters. There is no guarantee such measures will be enough to stop all long-term effects of acid rain on fishing rivers. More important, fishermen are not the only potential beneficiaries. Remember those forests, crops, buildings, animals and humans.

### The lesser of evils

Using less fossil fuel would reduce acid rain, but also increase pressures to replace thermal by nuclear or hydroelectric power stations. Both are *clean* from the current pollution viewpoint, but we need only link the words *Chernobyl* and *nuclear* to spell new doubts. Hydroelectric power has the superficial attraction that it uses non-polluting, radioactive-free water resources. But even when abundant water is available, to harness it for power generation is not without problems – building dams may destroy valuable agricultural lands, forests or magnificent scenery. A leading British environmentalist recently faced imprisonment for his physical actions to halt the construction of a dam, part of a proposed hydroelectric scheme that would have flooded unique forest scenery in an area of intense biological interest in the antipodean island of Tasmania. Vociferous international condemnation, highly-charged with emotion, killed that scheme. Scientists were deeply divided about its merits and dangers, highlighting the dilemmas of uncertainty.

### Sources of pollution

Major sources of atmospheric pollution in the U.K. are as follows:

(i) Power stations (over 60% of sulphur dioxide and nearly half the nitrogen oxides).

ii) Transport (nearly 90% of carbon monoxide, 40% of hydrocarbons and more than 25% of nitrogen oxides).

(iii) Chemical processes (nearly half the hydrocarbons).

U.K. power stations emitted between two and three *million* tons of sulphur dioxide per annum in the early 1980s. A lot of gas, but only about one-twelfth of the West European total. The Iron Curtain countries are no laggards when it comes to pollution, as Germany knows all too well when cold winter air from Siberia is sucked westward across the intervening industrial belt.

## The need for information

We need a lot more relevant information as a base for rational decisions on pollution control. To get and assess it requires collaboration between scientists, economists, politicians and diplomats. Data, once collected, often need careful statistical analysis.

*Waldsterben*, or forest decline, affects more than half the forests of West Germany. With one-third of their land timbered, it is no surprise the Germans are alarmed; the threat has been a stimulus for wide-ranging research to establish causes and find cures. Biologists, foresters, economists, industrialists, politicians and environmentalists are at the alert. The cause is *almost certainly* secondary effects of atmospheric pollution at international level.

The uncertainties are perplexing, danger levels difficult to quantify and preventive action hampered by the international dimension. Ozone, a secondary product from nitrogen oxides and hydrocarbons in the atmosphere, is almost certainly one contributor: it is known to damage trees at high concentrations. Yet *Waldsterben* is also found where ozone concentrations are low! Release of aluminium from soil is then high on the list of suspects.

Whatever the final cause (or, more likely, causes) of forest damage, tracking down the source of the initial pollutant in the chain needs a deeper understanding of weather patterns and the complex chemical reactions in the atmosphere (what comes down differing so often from what went up).

Early detection is yet another problem with pollution damage, particularly with slow growers like trees, for their growth rates vary from year to year. When affected by pollutants, the first effect may be that they grow more slowly. This may be attributed at the time, incorrectly, to a sequence of poor growing years linked to climatic conditions. From such climatic setbacks, trees normally recover; in Germany in the 1970s and 1980s they did not. In 1980 clearly visible signs of *Waldsterben* became widespread in German forests. Yet for some ten to fifteen years many of these trees had shown a slowing in growth, measured by diameter increments. It was only when they died or showed other obvious symptoms (e.g. loss of foliage) that the danger became clear.

With *hindsight* suspicion fell on the secondary effects of acid rain and particle pollutants as the cause of that slower growth, a slowing attributed earlier to vague meteorological factors.

In German forests in 1986 root damage that *might* be indirectly attributable to atmospheric pollution was found in trees that looked tolerably healthy. Trees in nearby forests were already dying from pollution. Only a close watch over the next few years will confirm if this root damage is a first symptom of more pollution evils.

Where acidification is a suspected cause of tree damage, rain may not always be the culprit. We must not forget nature. Planting a forest affects *local* soil and water acidity and, as already mentioned, natural biological processes alter acidity. The global balances of these natural processes may not apply locally. However, the widespread incidence of *Waldsterben* is strong evidence this is no local change. Symptoms are now appearing in British forests.

### Episodes

A current concern is *episodic* deposition. In some places extremely acid rain falls on just a few days of the year. This is associated primarily with meteorological conditions that affect both rate of transport of pollutants and chemical changes taking place during that transport. Just one day of *very* acid rain may be sufficient to kill all or many of the fish in a lake or river.

*Acid snow* has been elevated to an arch-villain role in producing sudden increases in acidity; these increases are often associated with quick thaws. *Dirty snow* gives rise to particularly high acidity on melting. The 'dirt' in this snow is particle pollutants which are converted to acid while in the snow; on thawing, such snow may be ten times as acid as polluted rain falling in the same area.

## Crop effects

Of the gaseous pollutants, nitrogen oxides and hydrocarbons produce ozone. In the U.S.A. it has been shown that the effects of excess ozone on crops is most deleterious in those regions with highest cropping potential at normal background ozone levels. Particularly bad is the effect on corn and soybeans in the Midwest. It has been asserted that Iowa soybean crop yields might increase by twenty-five to fifty million bushels per annum if ozone were reduced to normal background levels.

One small irony is that some pollutants may be beneficial (e.g. nitrogen, in the appropriate form) to forests. Acidity, though damaging leaves, may alter susceptibility to pests (which may be either beneficial or harmful), as well as altering soil chemistry.

Even in Scandinavia acid rain gets a few credit points for an increasing duck population. By killing fish, it has left an increased supply of food for these birds.

## Effects on structures

Acids corrode stone in buildings as well as metal fittings. High ozone levels damage paint and rubber.

Nitrogen oxides emission has been less widely studied than sulphur dioxide, though it is well known to increase acidity and trigger ozone formation. Both in the U.K. and in the U.S.A. vehicle exhausts are responsible for nitrogen oxides and hydrocarbon emissions. Pollution from this source is particularly important locally, especially in cities. Although some may be carried over long distances, this is more likely for nitrogen compounds emitted by the tall stacks of power stations or other industrial plants.

Use of low-lead petrols or modifications to engines or exhaust systems are key ways of reducing harmful emissions by vehicles. Here Britain drags her heels. Americans began using low-lead fuel some ten years ago; it is now freely available in Western Europe and Australia. In 1987 there were low-lead pumps at only a handful of U.K. filling stations, but it should be more widely available by the time you read this. In fairness, low-lead fuel is not problem-free: there are potential pollution problems of a different kind, but these should not be too difficult to solve.

## The CEGB attitude

The U.K. Central Electricity Generating Board (CEGB) has spent lavishly on research both into emission control measures and into the chemistry of pollution damage, though some of their effort has been criticized by independent scientists as spectacular and costly rather than cost effective. Some of its findings are inconclusive and the Board (and the British Government) were for long reluctant to admit its outpourings were a major source of acid rain that denuded Scandinavian lakes of fish.

As evidence of guilt mounted, they and others suggested the cheapest way to overcome acidity in lakes was by 'liming' them to neutralize the acid. This was tried in Scandinavia and elsewhere. It reduces lake acidity, but nobody has yet found a satisfactory way to lime fast-flowing rivers. Many fish spawn in these and high acid levels are lethal to the spawn. In addition, aluminium released from the soil by acids finds its way to limed lakes and triggers undesirable chemical reactions with the neutralizer; although here again the chemistry is complex and this problem not universal.

### The present position

The CEGB *Special Research Report No. 20 – Acid Rain*, issued in August 1987 (Anon., 1987), contains a wealth of readily digestible scientific information on pollution chemistry, but leaves unanswered vital questions about responsibility and controls; the truth is nobody (environmentalist or pollution producer) yet has firm answers.

An official admission that U.K. power stations were partly responsible for Scandinavia's acid rain was made in March 1986, and somewhat belatedly and reluctantly in September 1986 the U.K. agreed to spend £600 million over ten years on emission control. This was criticized by conservationists in Britain and throughout Europe,and also by a large body of impartial scientists, as too little and almost too late.

### Control options

Stopping pollution at source carries dangers. A switch to low-sulphur coal, even oil, may look attractive. The increased cost might be acceptable, but there would be knock-on effects: the change might spell doom to some coal mines, perhaps to all if the change was to oil. For some countries, it could mean dependence on imported fuel. Oil-fired power stations bring new pollution threats, the problem shifting from sulphur to nitrogen gases.

Filtering to reduce emissions may be costly. The 1986 British proposal is expected to reduce emissions by only about 14% and to increase power costs by 1.5%, but that is a trivial figure when set beside a 1987 governmental edict that power prices should rise by over three times that amount to satisfy political motives for obtaining a higher return on capital in an already profitable industry. Cheaper control methods may only shift pollution problems elsewhere. Washing coal prior to burning reduces sulphur compounds, but then fuel efficiency and disposal of the resulting sludge is a problem. Scrubbing sulphur out of waste gases after burning has attractions, but, again, how do we dispose of the polluted water? River purification boards, and river users generally, will not welcome it in our streams. In the long term it might also increase already serious ocean pollution.

## Who pays?

Much argument about who should pay to reduce pollution hinges on the fact that damage by *our* pollutants may not be to *our* rivers but more likely to somebody else's. With the predominant wind currents over Europe, U.K. power stations cause relatively few problems within the British Isles (except in parts of northern England and Scotland). We now accept they are major contributors to Western Europe's acid rain; small comfort we are not the sole offenders.

It is easy to argue that any pollution producer has a moral obligation to do something about it. Something, but what? The truth is all developed countries contribute to their own, and one another's, pollution. It is a political hot potato how much each should contribute costwise to the worthy cause of reducing *world* pollution. A corresponding reduction in percentage pollution levels in two countries is unlikely to cost the same in each, or to be equally effective. A switch from high- to low-sulphur coal may mean only marginal cost adjustments for one country; if the other has no low-sulphur coal, it may have to switch to more costly and less reliable sources of imported fuel. What if one country fails to take action? That by the other will be reduced in effectiveness, at worst leaving the active country only with an increased economic burden. Considerations like these contributed to, but did not excuse, the slow response of the British Government and electricity authorities to emission control.

## The club

International pressures, including moves by the E.E.C. and the U.N., to reduce emissions are having some impact in Western Europe. Driven harder by political and economic realities than by scientific logic, the interested parties have set up a so-called *30 per cent club*: its objective, reduction in all countries of sulphur dioxide emissions by 30% of the 1980 levels by the mid-1990s. While some twenty nations soon agreed to this, Britain (at least until it was browbeaten reluctantly into some action in 1986) contributed words rather than action. Its defence that its emission levels have already been reduced by some 37% since 1970 was no cause for pride. Indeed, it highlights one of the nonsenses of the concept of a fixed per cent reduction, for absolute levels of U.K. emissions are still amongst the highest in Europe. Remember, too, that prevailing westerly air-currents ensure a good deal of this emission finds its way to Scandinavia and other areas experiencing clear-cut acid rain problems.

The 30 per cent club is a typical compromise in a controversial situation, capitulation to the truth that politics cannot be ignored. Political factors have an unfortunate tendency to modify action that is justified by scientific evidence. Despite its virtues, the democratic electoral system is hardly conducive to politicians taking a long-term view.

What is needed is *higher* reductions from *above-average* offenders (but care is needed in defining what is meant by *above average* – remember our warnings about inappropriate averages in Chapter 3). That approach has been preached, rather than practised, in the U.S.A., where penalties have been proposed for those who are responsible for above-average pollution *for a given level of production*.

It has been argued that the atmosphere may not respond to partial reductions in pollutants, because the chemical processes that produce the more damaging secondary products such as ozone or acids are restricted by other factors. This may be true for some secondary pollutants, but there is convincing evidence that, where emissions of sulphur dioxide have been reduced, there is often a corresponding decrease in acidity in lakes and rivers.

### Our shame

Britain's long sojourn beneath a barely relevant percentage reduction umbrella has been little short of scandalous. Typical was the attitude taken at an international conference on atmospheric pollution problems in Geneva in September 1985 by a spokesman for the British Government. His vacillatory statement was worthy of the mythical Sir Humphrey of BBC 'Yes Minister' fame. He affirmed:

We believe that a broad attack on air pollution by sulphur dioxide, nitrogen dioxide, hydrocarbons and the secondary pollutants derived from them is appropriate. We believe that this must encompass all significant sources and this is the approach we have pursued. The proposed agreement on sulphur oxides now before this Executive Body deals only with part of this broad problem. My Government accepts as an aim of policy the reduction of emissions to air of both sulphur dioxide and nitrogen oxides by 30% from the 1980 levels. We hope to achieve this aim by the end of the 1990s. This is an aim of policy rather than a formal commitment.

Despite the case for further research, at the time that statement was made the case for immediate action was already strong. Perhaps the most comprehensive and unbiased report on the then current situation is given in *Report of the Acid Rain Enquiry* (Anon., 1985); several of the items there are updated, though from somewhat less impartial viewpoints, in Pearce's (1987) *Acid Rain* and the CEGB *Special Research Report No. 20* referred to on p. 137.

### Guilty

On 19 March 1986 came that grudging official admission that sulphur emissions from British power stations caused acid rain pollution in Norway. William Waldegrave, Minister of State at the Department of the Environment, in reply to Norwegian representations that Britain join the 30 per cent club, was reported in *The Times* the following day as stating, 'We were not able to move nearly as fast as they would like in using additional investment to bring emissions down faster than they would come down anyway.'

It comes as no surprise that this evasive reply from a politician who, on other occasions, has shown a realistic appreciation of environmental problems was disappointing, not only to the Norwegian Government, but to the European scientific community at large, and certainly to dedicated environmental lobbyists.

A later statement by Waldegrave indicated greater appreciation of environmental problems. In February 1987 he wrote in the Economic and Social Research Council *Newsletter*:

In the late eighties environmental issues are firmly on the political

agenda ... By stressing integration rather than conflict I believe
we will achieve the balance we seek – a balance between the
natural environment in which we live and the economic environment in which we earn our living.

## The North American situation

In North America a more homogeneous political system has made
it easier to collect information to provide a sound basis for action.
There is still an international dimension as Canada and the U.S.A.
are both involved.

A key question, as in Europe, is: Do benefits from reduced pollutants justify control costs?

Americans also have to contend with two major difficulties met
in Europe:

 (i) identification of sources (perhaps hundreds of kilometres dis-
     tant);
(ii) what is the true economic damage and how intolerable
     socially is pollution?

The need for a trade-off between production demands and protection of the environment is accepted, but as in Europe, individuals
and groups differ widely on where the balance should be struck.
As with many socio-economic problems, interested parties may
reach different conclusions – even if they start from the same scientific evidence – reflecting the essential interdisciplinary nature of
modern decision making.

Difficulty in deciding how reduced emission in one area will
affect deposition in another is basically a forecasting problem (see
Chapter 6, p. 109). Again, damage may be assessed differently in
scientific, in cultural or in economic terms, with differing implications about what control programme is justified. These problems
are as fundamental in America as in Europe.

### Canadian pressures

The Canadian Government was convinced in the early 1980s
that fossil fuel combustion in the U.S.A. was damaging its lakes

and streams, while admitting its own industry also contributed. It proposed a 50% reduction in sulphur dioxide emissions in the eastern states of both the U.S.A. and Canada. Because it is a more industrialized country, this was obviously going to cost more in the U.S.A. It could also be argued that, since pollution problems were more widespread in the U.S.A., their gain would also be greater.

We have already drawn attention to the crudity of a percentage reduction in discussing Europe's 30 per cent club. In parts of North America even a 50% reduction could still leave levels undesirably high; in others it might be quite unnecessary.

Whatever uncertainty surrounds effects of pollutants, rates of emission from various sources can be determined. Most comes from highly industrialized areas. On p. 140 we mentioned the more equitable approach to reductions suggested in the U.S.A.: first penalize industries that have emission rates *greater than average* for a given amount of production.

The U.S.A. aims to reduce sulphur dioxide emissions by some 8 million tonnes per annum by the mid-1990s.

### Our choice

Environmental damage is not always easy to cost in economic terms. What is the loss if a species of animal becomes extinct, if an area of scenic beauty is destroyed, if a recreational fishing river or lake is depleted of fish? Repercussions from any of these may be severe in economic terms if they are crucial to a country's tourist industry. Forest damage is not confined to timber loss: it may result in flooding or erosion, affecting much wider areas.

Unquestionably, atmospheric pollution damages the environment. The precise extent of damage and the exact cause of each type is often hard to quantify. We may act now to reduce drastically all or some pollutants. It will cost money. If we do nothing now, we shall certainly have to act (and pay) later. Whether the future cost, in purely economic terms, will be higher is an imponderable. The environmental damage will certainly be greater. By acting now we might take some unnecessary or not very effective measures that will be regretted with hindsight. The balance of opinion in most industrialized, and in all afflicted, countries favours considerable reduction of pollution during the next decade. Britain, at the time of writing, is still hesitant.

Not surprisingly, industrialists tend to take a conservative view about the need for reduction; extremists in the environmentalist lobby make unrealistic demands for zero pollution. The utilities (p. 70) for each party are clearly very different, and probably not well thought out.

## A review

What a sad saga of uncertainty is unfolded in this chapter after the success story of the Salk vaccine.

The key problems are two:

(i) We know too little about the chemistry of pollution and how it relates to meteorological and geological and biological factors.

(ii) On environmental issues utilities and subjective probabilities have an important role, but, because of lack of understanding of many of the factors involved in pollution damage, there are wide variations between different individual and group evaluations of these utilities and probabilities.

Classical statistics and objective probabilities have a part in assessing uncertainties about levels of specific pollutants, in forecasting future levels and in estimating how pollutants can be attributed to each source and the material damage done by each. However, deciding just how much environmental damage can be tolerated and at what cost must inevitably belong to the realm of subjective probabilities and utilities. Lack of information, or uncertainty about its relevance, does not alter risk – it only makes it more difficult to describe or quantify.

Perhaps the most hopeful sign at present is that the utilities of the different parties seem now to be closer than they were even twelve months ago. All parties at least agree there must be a sharing of responsibilities to reduce pollution. Formation of the

30 per cent club and the decision by the CEGB to do more filtering are both steps that reflect widespread acceptance of high utilities for restriction of environmental damage. That enlightened politicians are accepting this shift in emphasis is acknowledged in the quotation from William Waldegrave's statement on p. 141.

Valid criticism of the concept of percentage reductions reflects a wide gulf between what is desirable and what is politically possible. The scientists' traditional shunning of uncertainty has often made them reluctant to spell out the likely consequences of inactivity; many eminent ones have been too objective in putting only proven facts before politicians.

The slow pace of international cooperation also gives cause for concern. There is little evidence that the super-pollution producers of Eastern Europe are as concerned as those in Western Europe about their own (and Western Europe's) environment.

Fred Pearce's *Acid Rain*, the Edinburgh Conference report *The Acid Rain Enquiry* and the CEGB *Special Research Report No. 20* are all recommended for more background information. Kennedy (1986) gives a good scientific account of both the chemistry of pollution and the more general aspects of the nitrogen and sulphur cycles. Sprent (1987: Chap. 7) describes man's impact on the natural nitrogen cycle.

# CHAPTER 9

# HIGH-LEVEL DISTURBANCES

## Monitoring the atmosphere

Unlike Ronald G. Prinn, Professor of Meteorology at the Massachusetts Institute of Technology (widely known as MIT), you may never have heard of Cape Grim. In 1985 Prinn wrote:

> Such measurements . . . combine to make Cape Grim arguably the most comprehensive single surface atmospheric monitoring facility in the world.

This appears in his foreword to *Baseline 83–84*, a report of work during 1983–84 at Cape Grim Baseline Air Pollution Station, a small sophisticated laboratory on a west-facing headland at the edge of pioneer farmland in remote north-western Tasmania (Figure 9.1).

Cape Grim is Australia's link in the air pollution monitoring network of the World Meteorological Organization, a Geneva-based U.N. agency. There are many baseline stations in northern latitudes, few below the equator, and only that in Antarctica is at a greater southern latitude than Cape Grim.

Key objective of a *baseline* station, and hence its name, is monitoring levels and changes in levels of various atmospheric components in air *typical of that of a large segment of the globe not distorted by local sources of pollution which could influence readings.*

## Why Grim?

A glance at the map of the world shows why Cape Grim was

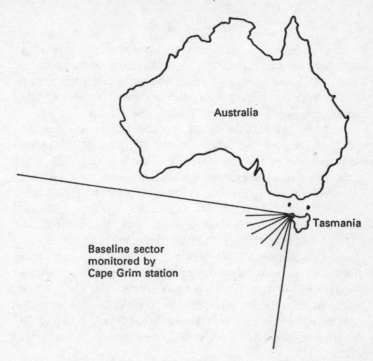

*Figure 9.1* Why Cape Grim is a suitable baseline station.

chosen. Westward, there is no pollution-producing land mass until one strikes the Argentinian coast of South America, more than halfway round the globe; Africa does not project south to Tasmanian latitudes. Antarctica is the only land of consequence to the south; rich in minerals maybe, but not yet industrialized. Thus, air reaching Tasmania from between west and south (the prevailing wind directions) should be as unpolluted as any: it is true *baseline air*. Even in such air, there are changes and trends for some atmospheric components owing to natural or man-made influences. In the last chapter we looked at the ground-level impact of air pollution. Of even greater concern, in the long term especially, is a universal build-up of pollutants and their secondary products in the global atmosphere.

### The baseline definition

*Baseline conditions* hold at Cape Grim only when the wind direction is between $190°$ and $280°$ (approximately between south and west), with speed at least 18 km/h, although that windspeed restriction is now relaxed for measuring some components. A further requirement is that, even given these conditions, there are no cyclic air currents drawing in pollution from the Australian mainland. There is an ingenious test for these. Such currents increase levels of the radioactive gas radon (see p. 10), a gas produced naturally over land masses, but not over the sea, so an increased radon level is a sure indicator that air is being cycled in from the Australian mainland. Confirmation often comes from tell-tale pollution blips in levels of several other components.

Data are automatically recorded for about two dozen components (many present only in parts per billion, a few in parts per trillion); for some, readings are made at a rate of 10 per minute! Hourly, daily or monthly means, as appropriate, are computer generated from these fundamental observations, together with a standard error (a measure of spread related to standard deviation); the latter enables us to get confidence intervals for each true mean.

Readings at Cape Grim and other baseline stations show erratic differences imposed on daily and seasonal variations, also long term trends. Carbon dioxide ($CO_2$) levels are increasing by 1 to 2 parts per million (ppm) each year; up from 330 ppm in 1976 to 345 ppm in 1986. Sophisticated measurements of the radioactive carbon component enable scientists to deduce how much of this is ambient or local change of a random nature and how much the product of biological processes or the burning of fossil fuels. $CO_2$ levels are important because they have a long-term influence on climate. It is predicted that doubling present $CO_2$ levels would increase mean global temperature by some $3°C$, the so-called *hothouse effect*, enough to degrade some farmland to desert and melt polar ice to submerge low coastlines.

*Table 9.1.* Mean baseline levels of ozone (ppbv), Cape Grim, 1982–84

| Month: | January | April | July | October |
|---|---|---|---|---|
| Mean level: | 14.9 | 24.6 | 32.0 | 27.8 |

## Atmospheric and stratospheric ozone

We turn our attention to ozone. We learnt on p. 136 that excess ozone is toxic to plants. In the upper atmosphere, ozone, although present in relatively small quantities, has a beneficial role as a filter of ultraviolet (u.v.) light, this radiation being a trigger for skin cancers.

Detecting long-term trends in ozone levels is a nightmare. There are large daily, and also seasonal, variations at any one spot and even bigger variations with latitude and altitude. Table 9.1 gives *average ground level* concentrations at Cape Grim under baseline conditions for the months of January, April, July and October between 1982 and 1984; the table is based on data reported by Douglas *et al.* (1986). The levels are parts per billion by volume (ppbv).

The July average is more than twice the January average. World-wide seasonal variations are minimal at the equator and increase at higher latitudes. Seasonal peak times differ with latitude and between the hemispheres.

Ozone near the ground is only a small proportion of the atmospheric total. Some 90% of it is in the stratosphere, 15–40 km above sea level, the concentration peaking at some 10 ppm at 20–25 km, more than 300 times those ground level concentrations of some 30 ppb measured at Cape Grim. A long-term ozone decline in the stratosphere would be disturbing. At the moment there is little evidence of a trend change in total ozone above given points on the earth's surface (the so-called column total), but we do know there are places where surface ozone levels are showing a steady increase. This suggests possible decreases in the stratosphere. Both changes are undesirable. Because it is harmful to crops (p. 136), we do not want increases at ground level; neither do we want decreases at higher levels, because of ozone's u.v. filtering properties.

### A theory of ozone depletion

In the journal *Nature*, Molina and Rowland (1974) put forward a theory that chemicals known as halocarbons or chlorofluorocarbons – CFCs in short – could reduce stratospheric ozone, allowing more u.v. light to reach the earth. The predicted consequences were as follows:

 (i)  an increase in skin cancer;
 (ii) changes in weather patterns;
(iii) effects on vegetation.

At the time there was no evidence to back the depletion theory, but CFCs were being used increasingly as aerosol propellants (about half their total use in the U.S.A.). They were, and still are, also used as refrigerants, solvents and for foam blowing. Based on the Molina and Rowland theory, the U.S. National Academy of Sciences produced a model indicating CFCs were likely to cause appreciable ozone depletion. The U.S. Government reacted quickly, banning non-essential use of CFCs as aerosol propellants in 1978, a decision made only after it seemed clear that alternative propellants would not bring new dangers. For some uses, however, there are limitations owing to fire risks with the alternatives.

The ban followed amendments to the Clean Air Act passed by Congress in 1977. This dealt specifically with 'human' modification of the stratosphere – especially the so-called ozone layer at 20–25 km. A statutory body, the Environmental Protection Agency, was asked to study effects of changes in the stratosphere on human health and crops and plant life; to monitor changes in the ozone level; and to examine methods of reducing release of CFCs and the economic effects of regulations aimed at reducing CFC release. Their study was to take due regard of feasibility and costs of proposed control measures.

Many countries responded positively to a Congress plea for international cooperation in research into stratospheric ozone levels, with less enthusiasm to a further request to adopt *standards and regulations consistent with those effective in the U.S.A.*

*Scientific action*

A scientific programme recommended by Congress was set in train at the time of the ban because compulsory reduction in propellant use of CFCs was being off-set by increases elsewhere. In 1982, predicted use by 1990 was 4.5 billion pounds weight per annum, up from 1 billion pounds at the time Molina and Rowland put forward their theory!

Using the theoretical model, the National Academy of Sciences predicted that, even if CFC emissions stayed at 1977 levels, ozone depletion in the next decade would be between 5 and 28% at critical stratospheric levels. The wide range reflected elements of uncertainty both in the mathematical equations and about the precise chemical processes involved. The model used probabilistic ideas, some of the probabilities doubtless more subjective than objective.

The Academy also came up with the complicated assertion that, if CFC release increased at the rate of 7% per annum, there was a 70% chance that ozone depletion would exceed 30% by the year 2000, a somewhat contorted way of saying that, with a 7% growth rate, there was likely to be high depletion.

*The Molina–Rowland theory*

The basics of the Molina and Rowland CFC–ozone theory were as follows:

(i) CFCs are virtually indestructible in the lower atmosphere (the troposphere), and thus accumulate.
(ii) They are known to be transported eventually to the stratosphere.
(iii) Laboratory experiments indicated that chlorine released from the CFCs would reduce the ozone.

These were hard facts. From the laboratory findings, they hypothesized a decrease could be expected in the stratosphere. Change in ozone levels with season, latitude and altitude makes this difficult to detect. This is where statisticians found a role using the means of forecasting known as time-series analysis. In 1982 Geoffrey S. Watson of Princeton University published a paper on this approach (Watson, 1982) and other developments were reported by Tiao (1983).

## Some doubts

Hard data soon cast doubts upon the Molina–Rowland theory, or at least suggested counter-balancing changes due to other factors. With little evidence of atmospheric depletion, industry, with a fairly characteristic response, resisted controls on use of CFCs, arguing it would be time enough to act when we had firm evidence. They received further consolation in 1981 when more refined theory, taking other factors into account, suggested only a 5–8% depletion likely.

But there was no room for complacency. A weakness in the industrialists' argument was that CFC pollutants take many years to reach the stratosphere. When they do, they are expected to stay there a long time: they are the exception (p. 131) to the Newtonian rule that what goes up must come down. After any ozone depletion, there will be a further lag of decades before we see an increased incidence of skin cancer, as this normally develops some thirty years after the vital exposure. Sunburn in childhood is often the trigger for skin cancers in middle age.

The Academy estimated that a 16% ozone depletion would lead to 300 000 additional cases of skin cancer of a non-fatal kind and 3000 additional fatal cancers per annum. Another prediction was that the percentage increase in skin cancer would be about four times any percentage decrease in ozone.

These predictions were based on incidence of skin cancer at various places and the known levels of u.v. radiation at each. Generally speaking, rates are higher in the tropics, where there is already insufficient ozone in the stratosphere to absorb the high levels of u.v. light. Tropical regions of Australia, where the u.v. component of sunlight is especially high, have some of the top skin cancer rates in the world.

## A complex process

We are gradually learning the chemistry of ozone depletion – a complex process involving temperature and atmospheric compositional changes at various altitudes. These changes influence photochemical reactions triggered by solar radiation.

Remember, maximum concentrations of ozone are only about 10 ppm. Indeed, if all the ozone in the atmospheric column above a given ground point were concentrated in one layer at the foot of the column, it would only be about 0.3 cm high, the exact height varying slightly with temperature and pressure changes.

Ironically, u.v. light actually produces absorbing ozone photochemically in the stratosphere. The balance is maintained by infrared (i.r.) light in turn breaking this ozone down to oxygen. Other factors, including $CO_2$ levels, also help maintain the balance.

Intrusion of chlorine or hydrogen upsets the natural balance. Small amounts of these (of the order of parts per billion) released by CFCs have a catalytic effect that breaks down stratospheric ozone. A catalyst is a substance that triggers a chemical reaction while itself remaining chemically unchanged.

There are complementary ozone-producing catalysts, including laughing gas (nitrous oxide) and other oxides of nitrogen in exhaust gases from jet aircraft. Nitrous oxide is also released by natural biological processes, but levels in the atmosphere are balanced over long periods. Nitrogen oxides play a complex role in the ozone story, increasing ozone near the surface (p. 136), but, when pumped into the stratosphere by high-flying jets, the catalytic effect is complex as the chemicals form a 'smog' with hydrocarbons, a smog that has a temperature effect that may influence ozone production. We mentioned (p. 148) increases in baseline carbon dioxide levels. $CO_2$ does not affect ozone levels directly, but we must not forget that an increase raises temperature and that affects the photochemical processes. The effect of carbon monoxide and some other pollutants is small and uncertain, but may lead to ozone increase: bad at ground level; good in the stratosphere.

Fortuitously, for some time, these factors seem to have been broadly in balance. For how long will depend critically upon the relative levels of the CFCs and aircraft exhaust gases we pump into the stratosphere. There are growing fears (and evidence) that the balance may be getting upset.

## Measurement techniques

Measuring level of CFCs, nitrogen oxides or ozone at high altitudes and at different times is not easy. Perturbations in concentration when looking at parts per million, billion or even trillion are even harder to detect. Rockets, balloons and aircraft are used to collect stratospheric samples. Help comes too from a ground network – the Dobson stations. These record regularly the total ozone column above the points at which they are situated. Some can even estimate the concentrations at various heights in the column, estimates that are checked periodically against those samples collected by aircraft, balloons, etc. Data are coordinated by the World Meteorological Organization (which also coordinates baseline data) and recorded in books familiar to scientists studying ozone as the 'red books'. In keeping with the computer age, the data are also available on magnetic tapes. A frustrating constraint on building a composite picture is the uneven siting of Dobson stations: most are in North America, Western Europe, India and Australia.

## A problem of detail

We can explain broad features of ozone variation, but not fine detail. Time-series analysis looks promising, but is restricted by limitations in data. We need relevant information over a long period to separate out random fluctuations and seasonal or positional variations that are imposed on any trend. Dobson stations have operated for over fifty years, but only since about 1958 (the International Geophysical Year) has data of sufficient accuracy to detect trends been collected. Even these data may not suffice to detect ozone depletion at critical levels before it is too late, a concern reinforced by the time lag in CFCs reaching the stratosphere. The global picture is not adequately reflected by Dobson data because of those large areas (Africa, South America, much of Asia) with few or no stations.

On top of these data limitations, detailed analysis of ozone changes must also take into account the effects of solar activity, atomic explosions and volcanic eruptions.

## The Antarctic hole

Scientists were alarmed late in 1986 when a 'hole' appeared in the stratospheric ozone layer above Antarctica, normally a fairly high-ozone region. This was not a completely new phenomenon, more an acceleration in one first noticed some ten years ago, and since observed annually each September and October. An ozone depletion from about 2 to 0.5 ppm took place between 17 and 19 km altitude from August to October 1986 and there was also a general depletion between 12 and 20 km altitude. Temperature changes in the stratosphere and a reduction in the polar stratospheric cloud were also evident. This was a classic chicken and egg situation – were the temperature changes caused by ozone depletion or vice versa? One theory is that evaporation of the cloud may have released chlorine or nitrogenous catalysts (secondary products of pollution) to cause the sudden depletion. A spaceflight to probe this hole and new methods of detecting trace elements in the stratosphere were used in a 1987 study. Preliminary findings from that study suggested the 'hole' was larger (in geographic area) than in previous years. By the time you read this, more should be known about what might be an ominous threat.

## Theoretical and practical uncertainty

The ozone story is an example of uncertainty compounded because observation has not caught up with theoretical models.

The theoretical *expert guess* at first seemed to overestimate depletion rates, but this was because it did not take account of balancing factors. If these factors change, balance may be lost. Then, by the time we collect data, it might be too late to avert danger.

In mid-1987 the U.S.A. was pressing for a 95% reduction in CFC use over fourteen years; the EEC wanted a freeze for two years at 1986 levels, followed by a 20% reduction. Some environmentalists claim an 85% reduction is needed to confine ozone depletion to its present rate (which is not well determined). This confusing picture represents different utilities reflecting the cost to industry, values placed on health and even selfishness, the differences in part only being a consequence of inadequacies in information.

## Alternative strategies

Alternative strategies are possible in many risk situations. We could make a once and for all decision (ban CFCs and/or high-flying aircraft). Most people would regard this as too drastic. The alternative is to adopt an adaptive strategy of making changes as evidence builds up. This carries a danger that we may too readily discount the long transport times for CFCs to the stratosphere, and the stability of the catalysts released when they get there, so action may be too late.

We must decide how much extra we, as a world society, are prepared to pay for the products we want (or what will we sacrifice) to avoid an increased risk of a disfiguring but non-fatal skin cancer and a smaller increased chance of a fatal skin cancer, knowing it may be a danger to future generations, not our own.

Optimists may argue there are good prospects that a cure for skin cancer will be available by the time the threat becomes a reality. Some of us may think such a cure will be available within the next decade. We may be inclined to assign our own personal or subjective probability to such being the case, but even among rational people there is likely to be a wide range of numerical values for such a probability. The supreme optimist may think it worth a better than even-money bet there will be a cure; the pessimist offers 100 to 1 against. In opinion poll terminology the wise might well stay in the *don't know* category. We must not forget effects on climate and crops of changing ozone levels.

## A review

Thanks to the U.S. Government collaborating with an impartial body, the U.S. Academy of Sciences, to encourage research into ozone depletion in the stratosphere, more scientific progress has been made with this than with most atmospheric pollution problems. Uncertainties that are within the domain of classical objective probability and statistics are largely concerned with difficulties of measurement and in prediction by time-series analysis.

There is a substantial amount of data and a reasonable understanding of the chemistry of the processes involved, although the uneven distribution of recording stations and difficulties of measurement in the stratosphere leave important data gaps.

The risk element of the problem inevitably involves subjective probabilities and utilities. There is general agreement that depletion of stratospheric ozone (because of its role as a u.v. filter and other secondary effects on crops and weather) is undesirable. Controversy is mainly about whether the CFCs are the prime cause of that depletion. The claim by industry that we should wait and see is dangerous because of the time lag between cause and effect on the ozone layer and the delay between initiation of skin cancers and the appearance of symptoms.

While the U.S.A. has taken firm legislative action to reduce the use of CFCs, a lack of international agreement has meant that their total use is still increasing, emphasizing the urgency of better international regulatory controls of a dangerous group of chemicals.

There are signs the EEC is shifting towards the U.S. attitude on controls; sadly the U.K. is among the current heel draggers.

---

# CHAPTER 10

# OILING THE WHEELS OF INDUSTRY

## Down to earth

After a voyage of unpredictability to a polluted stratosphere, we return to earth to look at more mundane manifestations of uncertainty, situations where we use objective probabilities and well-established statistical methods. We move from dramatic risks associated with pollution to humdrum and unavoidable elements of variability on the shop floor and in the marketplace. Here, a quantitative approach to solving problems has contributed strikingly to many post-war industrial success stories. The emphasis is on the search for high quality and greater efficiency and cost-effective processes.

This chapter relies more heavily than others on the content of Chapters 4 to 6. If you only skimmed through these, it may be best to do the same for this chapter at first reading.

## A new technology

Born of World War II, a new technology blossomed in the 1950s and 1960s and is now well established. Dubbed *operational research* in the U.K., America preferred *operations research*, universally abbreviated to O.R.

There is no simple or precise definition of O.R. Basically, it is the use of mathematical methods to guide management to better policies and optimal decisions; some of it is applied statistics masquerading under a more pompous name. There are also non-statistical aspects. For example, we may seek the best of a number of clearly defined alternatives to minimize production costs or maximize profit; linear and dynamic programming techniques are well-known methods in this context.

O.R. problems often lead to intractable mathematics. Its development owes much to computers, reducing computational time from years to microseconds and, even more importantly, giving a way to test in minutes what might happen in real situations over weeks, even years, if we decide to take certain actions.

One technique for doing this is *simulation*. We shall be illustrating the principle by almost trivial examples – trivial in the sense that there exist simple and exact mathematical solutions to our problems. This simplicity enables us to demonstrate the practical validity of simulation results.

### Uncertainty and pattern

Short-term uncertainty but long-term pattern is a familiar phenomenon. The numbers of customers entering a bank or a post office varies from hour to hour and from day to day, but totals per week or per month or per year either remain relatively constant, or perhaps show some trend or steady recurrent feature (e.g. higher on Friday than on Wednesdays, more in summer than winter, increasing from year to year, etc.).

Some machines have components that wear quickly and need frequent replacement. We seldom know *exactly* how long a machine will operate before it breaks down, but for a large batch of similar machines we may find that failures nearly always occur after the machine has run, say, between 9 and 16 hours.

The statistician sets up a mathematical model to mirror the *distribution* of times to failure. This can be used to make predictions about the behaviour of the system, forecasting such things as the long-run percentage of failure times greater than 12 hours or the probability there will be a failure in less than 14 hours or the average time to failure.

If the mathematical model is too complex to manipulate analytically, we may still get approximate answers by simulation. We still need the model, but the speed of the computer gives quick answers that take weeks, even years, to get simply by observing machines in operation, always provided our model reflects reality. Even when there is an exact mathematical solution, simulation may still be useful if it saves extensive algebra.

### A simple example

A machine has three components A, B and C. It breaks down if and only if one or more of these fails. Experience has shown the following:

(i) Component A is equally likely to operate 10, 11, 12, 13, 14, 15, 16, 17, 18 or 19 hours.

(ii) Component B is equally likely to operate 10, 11, 12 or 13 hours, but twice as likely to last 14, 15 or 16 hours.

(iii) Component C is equally likely to operate either 9 or 12 hours, but twice as likely to last 13, 14, 15 or 16 hours.

(iv) Breakdowns in the various components are independent events.

This experience (experimental evidence) about component failure is the basis of a mathematical model to examine machine behaviour.

This example looks artificial. Why should a component last an exact number of hours? Why not 15 hours 43 minutes or 13 hours 11 minutes? We expect this in real life. But our model is realistic if an inspection is made hourly and at each such inspection it is possible to predict whether a component is very likely to break down within the next hour (e.g. a component may become hot or change shape an hour or so before failure). If replacement is made whenever such a symptom is detected at inspection, we, in effect, record this as the 'time of failure'.

We make one more assumption. If a potential failure is detected in one component, then all three are replaced. This is not unreasonable if stopping the machine for replacement is costly (e.g. in lost production time) compared with the price of components, for it is evident with this policy that when one component fails the others are also likely to be approaching the end of their useful life.

*Table 10.1.* Distribution of breakdown times

| No. of hours: | 9 | 10 | 11 | 12 | 13 | 14 | 15 | 16 |
|---|---|---|---|---|---|---|---|---|
| Probability: | 0.100 | 0.171 | 0.153 | 0.184 | 0.176 | 0.136 | 0.064 | 0.016 |

Simulation has two main uses in a situation like this:

(i) It shows quickly how a system based on given assumptions will work in practice.

(ii) With little extra effort, we may study how changing assumptions alters system behaviour.

The second use is often the more important. If a rival manufacturer offers an alternative to component A that always lasts, each with equal probability, either 12, 13, 14, 15 or 16 hours, we could replace the original information for component A by this new set of probabilities and find out how our new system would behave and see if there is improved performance. Given the relative costs of the old and new component A, we can decide if a change would be cost effective.

### Practical details

We consider first simulations with the original components A, B and C. A glance at our information (p. 160) shows the *minimum* number of hours the machine will operate without a breakdown is 9 and the maximum, 16. Agreed? This is because all components last *at least* 9 hours, but two components *never* exceed a life of 16 hours.

It is less easy to see how often we *expect* (i.e. what is the long-term relative frequency, or probability, of) a breakdown after each of 9, 10, 11, 12, ..., 16 hours. With careful thought plus tedious arithmetic, we may calculate the probability that one or more component fails at 13 (or any other number of) hours, *conditional* upon no earlier component failure. Table 10.1 gives these probabilities, relevant to the distribution of the random variable *number of hours to failure*. By all means, try the tedious exercise of calculating the probabilities if you wish; if not, please take them on trust. Assumption (iv) (p. 160) that all breakdowns are independent is vital to these calculations.

*Table 10.2.* Relative frequency of times to breakdown in 2000 simulations

| No. of hours: | 9 | 10 | 11 | 12 | 13 | 14 | 15 | 16 |
|---|---|---|---|---|---|---|---|---|
| Rel. freq.: | 0.092 | 0.178 | 0.157 | 0.189 | 0.151 | 0.142 | 0.070 | 0.021 |

We may calculate the mean, or expected, time to breakdown from Table 10.1 using the methods on p. 83. That is, we multiply each lifetime by its probability and add:

$$E(X) = 9 \times 0.100 + 10 \times 0.171$$
$$+ \cdots + 16 \times 0.016 = 11.909.$$

With a computer we may simulate what happens in practice, calculating times to breakdown on a large number of occasions. The limiting relative frequency concept of probability then gives us approximations to the values in Table 10.1. Here, use of a simulation model is trivial because we already know the answer, but it is an interesting exercise and one that is easily programmed in BASIC (or a more sophisticated language) for a home or office computer. Table 10.2 gives the observed relative frequency of breakdowns for 2000 simulations.

Do you think these frequencies show reasonable agreement with the probabilities in Table 10.1? The largest discrepancy occurs for 13 hours. We estimate the expected time to breakdown by replacing true probabilities by the relative frequency estimates. It is 11.94, close to the theoretical expectation of 11.909 calculated above.

### The mechanics of simulation

A basic tool for simulation is a set of random digits (p. 109). Most computers have software to generate these. Remember (p. 108) two key properties of random digits:

(i) Each digit is equally likely.
(ii) Foreknowledge of digits that have already occurred give no indication of those that will occur next in the sequence.

One warning: some random digit generators are not as reliable as we would wish, the probability for each digit not being exactly

0.1 as it should be. Most modern computers do a reasonable job at random number generation, but it might be worth testing output from one you use regularly with a test for equal frequencies given in *Understanding Data* (Sprent, 1988: p. 220) or in most elementary statistics text books. More elaborate tests are needed to expose other possible weaknesses of random number generators.

## Using random digits

Here is how we use random digits in this example. We are told (p. 160) component A is equally likely to break down at any of ten times 10, 11, ..., 19 hours, so we associate a probability of 0.1 with each time. This is also the probability any particular digit occurs at any point in a sequence of random digits. So we *simulate* the pattern of breakdown times for component A by associating one *specified* digit with each breakdown time. An obvious thing to do is associate 0 with 10 hours, 1 with 11, 2 with 12, and so on, finally associating 9 with 19 hours; for example, the random digit 5 implies component A breaks down at 15 hours.

For component B, our statement (p. 160) that it is equally likely to last any of 10, 11, 12 or 13 hours, but twice as likely to last 14, 15 or 16 hours, leads to the conclusion that the probability of a life of 10, 11, 12 or 13 hours is in each case 0.1, while the probability of a life of 14, 15 or 16 hours is in each case 0.2. The sum of these probabilities over all lifetimers is unity, as it must be for a mutually exclusive and exhaustive set of outcomes. To simulate, we associate a life of 10, 11, 12 or 13 hours each with one digit, say 0 for 10, 1 for 11, 2 for 12, 3 for 13; we associate two digits (4 and 5) with 14, two more (6 and 7) with 15, then 8 and 9 with 16 hours, giving a probability of 0.2 for each of 14, 15 or 16 hours.

Similar reasoning (be sure you see why) suggests that for component C we equate digit 0 with 9 hours, 1 with 12, 2 and 3 with 13, 4 and 5 with 14, 6 and 7 with 15, and 8 and 9 with 16 hours to give appropriate probabilities.

These associations between failure times and random numbers are summarized in Table 10.3.

*Table 10.3.* Digits specifying time to breakdown of three components

| Breakdown time (h): | 9 | 10 | 11 | 12 | 13 | 14 | 15 | 16 | 17 | 18 | 19 |
|---|---|---|---|---|---|---|---|---|---|---|---|
| Component A: | | 0 | 1 | 2 | 3 | 4 | 5 | 6 | 7 | 8 | 9 |
| Component B: | | 0 | 1 | 2 | 3 | 4 | 6 | 8 | | | |
| | | | | | 5 | 7 | 9 | | | | |
| Component C: | 0 | | | 1 | 2 | 4 | 6 | 8 | | | |
| | | | | | 3 | 5 | 7 | 9 | | | |

*Table 10.4.* Random digits

```
13705 24898 41152 98039 46952 91003 42856 79735 35746 47627
61410 46055 22008 71066 44846 79008 92992 19182 56667 64522
74964 42633 11452 20678 64959 44196 81631 33737 27369 53363
03935 29613 22583 01829 23355 97940 11693 19898 30068 91553
13086 34689 16723 97914 26871 52922 94525 28217 08565 26720
```

For each simulation our program needs three random digits, one for each component. We use the equivalences in Table 10.3 to get a breakdown time for each component. The time to machine failure is determined by the *earliest* of these. Thus, the three digits 7, 9 and 4 would imply (Table 10.3) that component A is going to break down after 17 hours (digit 7), component B after 16 hours (digit 9) and component C after 14 hours (digit 4). So component C is going to break down first at 14 hours. Since all components are replaced when one fails, the times to failure for A and B now become of academic interest as the machine is restarted with three new components.

The computer selects three more random digits and evaluates a new time to machine failure. A fast and well-programmed computer will complete many thousand simulations in a matter of seconds.

One may simulate with pencil and paper the procedure for a few simulations using published tables of random digits (see, for example, Neave, 1981: p. 42). Table 10.4 gives a short run of such digits for use in the examples in this chapter.

Table 10.5 gives results for twenty simulations using random

*Table 10.5* A simulation of times to machine failure

| Simulation number | Random digits | A | Life B | C | Time to machine failure |
|---|---|---|---|---|---|
| 1 | 2 4 8 | 12 | 14 | 16 | 12 |
| 2 | 9 8 4 | 19 | 16 | 14 | 14 |
| 3 | 1 1 5 | 11 | 11 | 14 | 11 |
| 4 | 2 9 8 | 12 | 16 | 16 | 12 |
| 5 | 0 3 9 | 10 | 13 | 16 | 10 |
| 6 | 4 6 9 | 14 | 15 | 16 | 14 |
| 7 | 5 2 9 | 15 | 12 | 16 | 12 |
| 8 | 1 0 0 | 11 | 10 | 9 | 9 |
| 9 | 3 4 2 | 13 | 14 | 13 | 13 |
| 10 | 8 5 6 | 18 | 14 | 15 | 14 |
| 11 | 7 9 7 | 17 | 16 | 15 | 15 |
| 12 | 3 5 3 | 13 | 14 | 13 | 13 |
| 13 | 5 7 4 | 15 | 15 | 14 | 14 |
| 14 | 6 4 7 | 15 | 14 | 15 | 14 |
| 15 | 6 2 7 | 16 | 12 | 15 | 12 |
| 16 | 6 1 4 | 16 | 11 | 14 | 11 |
| 17 | 1 0 4 | 11 | 10 | 14 | 10 |
| 18 | 6 0 5 | 16 | 10 | 14 | 10 |
| 19 | 5 2 2 | 15 | 12 | 13 | 12 |
| 20 | 0 0 8 | 10 | 10 | 16 | 10 |
| | | | Mean time to failure | | 12.1 |

digits selected from Table 10.4, starting arbitrarily with the second group of five in the first row selecting the triplet 248, 984 and 115, and so on. We must always enter a table initially at some arbitrarily selected and different point. If we always started at the beginning, everybody would get the same simulations. Even for these twenty simulations, the mean failure time (total failure time for all simulations divided by the number of simulations) is 12.1, reasonably close to the theoretical expected value of 11.909. Of course, with just a few observations, the relative frequencies of breakdowns of each duration could not be expected to mirror closely the probabilities in Table 10.1.

*Table 10.6.* Probabilities of various times to failure of three components

| Time to failure (h): | 11 | 12 | 13 | 14 | 15 | 16 |
|---|---|---|---|---|---|---|
| Component A: | | 0.2 | 0.2 | 0.2 | 0.2 | 0.2 |
| Component B: | 0.1 | 0.2 | 0.2 | 0.2 | 0.2 | 0.1 |
| Component C: | | 0.3 | 0.3 | 0.2 | 0.2 | |

### *Altering the system – is it economical?*

Suppose the cost of replacing all three components, including the lost time element, when the machine fails is £61 but an alternative supplier offers a set of components that are more expensive and result in a total cost of £63. He defends the extra cost by claiming a longer average interval between failures. The buyer is not convinced, so he asks for, and is given, the lifetime distributions for individual components. He is told that component A is equally likely to last exactly 12, 13, 14, 15 or 16 hours; component B is equally likely to last any of 12, 13, 14 or 15 hours, but only half as likely to last either 11 or 16 hours; component C is equally likely to last 12 or 13 hours, and two-thirds as likely to last 14 or 15 hours.

Given this information, it is easy to work out the probabilities of failures at each time for each component. These are given in Table 10.6.

To simulate, we now assign, as we did in Table 10.3, random digits to reflect the probabilities. An appropriate set is given in Table 10.7.

With this random digit allocation, our simulation program is used to estimate probabilities (limiting relative frequencies of each possible time to failure).

A set of 1000 simulations give the relative frequencies in Table 10.8 (e.g. the frequency 0.283 for 13 hours implies that in 283 out of 1000 simulations the time to first component failure was 13 hours).

The estimated expected mean time to breakdown is 12.404 hours. With the old components our estimate (based on 2000

*Table 10.7.* Random digit allocation reflecting probabilities in Table 9.5

| Time to failure (h): | 11 | 12 | 13 | 14 | 15 | 16 |
|---|---|---|---|---|---|---|
| Component A: | | 0 | 2 | 4 | 6 | 8 |
| | | 1 | 3 | 5 | 7 | 9 |
| Component B: | 0 | 1 | 3 | 5 | 7 | 9 |
| | | 2 | 4 | 6 | 8 | |
| Component C: | | 0 | 3 | 6 | 8 | |
| | | 1 | 4 | 7 | 9 | |
| | | 2 | 5 | | | |

*Table 10.8.* Relative frequency of breakdown times in 1000 simulations

| Hours to failure: | 11 | 12 | 13 | 14 | 15 |
|---|---|---|---|---|---|
| Rel. freq.: | 0.102 | 0.513 | 0.283 | 0.083 | 0.019 |

simulations) was 11.94 hours. Is it worth the additional £2 – the increase from £61 to £63 – for this improvement?

## True or false?

See if you agree with this argument. The fairest comparison is between average costs per unit time (e.g. per hour). A cost of £61 for a replacement *on average* every 11.94 hours represents an hourly cost of £(61/11.94) = £5.11 (to the nearest penny). Now, a cost of £63 for a replacement *on average* every 12.404 hours represents an hourly cost of £(63/12.404) = £5.08. Thus, there is a small saving for the new components. This conclusion is based only on simulation studies: in one case, 2000 simulations; in the other, only 1000. Table 10.1 gave exact probabilities for the first types of component, indicating a true mean of 11.909. We can also work out exact probabilities and a true mean using the new type of component since we know the lifetime distributions for each. These probabilities are given in Table 10.9.

Note there is reasonable agreement with the probabilities

*Table 10.9.* Probabilities of times to breakdown with new components

| Hours: | 11 | 12 | 13 | 14 | 15 |
|---|---|---|---|---|---|
| Probability: | 0.100 | 0.508 | 0.272 | 0.096 | 0.024 |

estimated by simulations in Table 10.8. Using Table 10.9, we find the true mean is 12.436.

For the two types of components, we now calculate the true average costs per hour as £(61/11.909) = £5.12 for the old and £(63/12.436) = £5.07 for the new, again suggesting a small saving, our estimate being a slightly higher saving using the true values than that obtained from simulation.

### Have we played fair?

The above arguments look reasonable, but think back to some of our examples about means in Chapter 3. Are we not perhaps taking an invalid short-cut like the one that led to an erroneous estimate of average speed in Example 3.8?

Indeed we are. What we must do is work out a *cost per hour* if breakdown occurs at each possible time. We know the probability of each such cost arising. It is simply the probability of a breakdown at that time, so we can work out a mean or expected cost per hour. To illustrate the procedure, take the case of the probabilities given in Table 10.9. The renewal cost per hour in pounds if the machine runs 11 hours is 63/11 = 5.727. Similarly, the hourly renewal costs, in pounds, for 12, 13, 14 or 15 hours are, respectively, 63/12, 63/13, 63/14, 63/15. If we multiply each of these by the relevant probabilities given in Table 10.9 and add, we get the *expected hourly renewal cost*, which turns out to be £5.09. A similar procedure using the times and probabilities in Table 10.1 with a replacement cost of £61 gives an expected hourly renewal cost of £5.24, implying an average saving of 15p per hour by switching to the new components. Similar calculations using the simulated estimates in Tables 10.8 and 10.2, respectively, give average hourly costs of £5.10 and £5.23, very close to the true values.

*Table 10.10.* Distribution of breakdown times for original component B

| Hours: | 10 | 11 | 12 | 13 | 14 | 15 | 16 |
|---|---|---|---|---|---|---|---|
| Probability: | 0.1 | 0.1 | 0.1 | 0.1 | 0.2 | 0.2 | 0.2 |

*Table 10.11.* Distribution of breakdown times for new component B

| Hours: | 11 | 12 | 13 | 14 | 15 | 16 |
|---|---|---|---|---|---|---|
| Probability: | 0.1 | 0.2 | 0.2 | 0.2 | 0.2 | 0.1 |

We see that in this example simulation results are close to the known true ones, but simulation really comes into its own in more complex situations.

## Can worse be better?

As another check on intuition, we look more closely at our last example. It seems intuitively obvious that, if we increase the average life of all components, we should increase the expected time to breakdown, whereas, if we decrease the average lives, we will reduce the expected time to breakdown.

For each component, we are given a distribution of the time it lasts, so we can work out an average lifetime. For the original component B, the probabilities (p. 160) are given in Table 10.10.

From the table, we easily calculate the expected time to breakdown as

$$10 \times 0.1 + 11 \times 0.1 + \cdots + 16 \times 0.2 = 13.6.$$

For the new component B, the distribution is given in Table 10.11.

The expected time to breakdown is now (check it) 13.5; that is, less than that for the original component B. Yet we found the machine runs on average longer with the new components. Perhaps this is because the new components A or C do better *on average* than their old counterparts. The expected times to breakdown for all original and new components are given in Table 10.12.

So each new-type component has a shorter life expectation

*Table 10.12.* Expected times to breakdown for old and new components

| Component | Original | New |
|-----------|----------|-----|
| A | 14.5 | 14.0 |
| B | 13.6 | 13.5 |
| C | 13.7 | 13.3 |

than the old, yet our simulations showed that with the new components the expected time to machine breakdown is longer!

So what about intuition? Do you see where it is letting us down? The point is that the machine breaks down at the *first* component failure. Clearly, in the nature of random variation, in many simulations some (perhaps all) components are likely to break down at less than the *expected* (average) time. Thus, we might anticipate the average machine run-time will be less than the average life for individual components. This is so.

With the original components, the expected life was 11.909 hours; with the new components, 12.436. Both are less than the means for any single component. Clearly, if components have a greater *spread* of breakdown times, there is an enhanced probability of a (relatively) earlier breakdown. With the original components, a breakdown at 9 hours was possible; with the new, 11 hours is the minimum time to breakdown. Looking at Tables 10.3 and 10.7, we see that the range of possible breakdown times for each of the original components (Table 10.3) is greater than that for each corresponding new component (Table 10.7). That explains why we get a longer expected life with the new components. Another way of putting it is to say that the new components are more consistent (i.e. less variable) than the old.

## Distribution of extremes

Consistency being more important than means in determining time to *first* component failure draws attention to an important kind of uncertainty – that associated with extremes such as the earliest or latest times an event occurs. We may likewise be interested in the greatest height of a tide or the fastest time in which a race may be run.

Extremes are important to engineers in matters like flood control or stresses on a bridge or building. To design a dam to control flood waters (perhaps with secondary uses for irrigation or hydro-electric power), it is important to have a reasonable estimate of the greatest flood ever likely to occur. Plans for a bridge require estimates of the greatest stress that might occur, both from traffic loads and from winds or flood waters. Many nineteenth-century structures were overdesigned because of a rule of thumb once used: *calculate the greatest foreseeable stress; for safety, build to stand double this.* A conservative rule, but one that did not always work. The original Tay railway bridge failed in 1879 partly because specifications were not adhered to, more fundamentally because calculations did not allow appropriately for stresses due to wind – the bridge was too rigid! This was the scientific explanation, though some of the more puritanical churchmen of the day ascribed the accident to a Heavenly condemnation of those who travelled on the Sabbath.

Estimation of extremes is an important branch of statistics. Until Roger Bannister ran a four-minute mile in 1954 (breaking a record that had stood for nine years), many people regarded this as an impossible target. Will we ever see a 'three-minute mile'? That Bannister's record was soon broken suggests a psychological barrier associated with four minutes.

## Queues

Simulation is also useful to study how queues behave. Again we take a simple situation where we can test our finding against easily calculated results. Suppose it is known that people join a queue either singly with probability 0.8, in pairs with probability 0.1 or in threes with probability 0.1. The expected number of persons per arrival is then $1 \times 0.8 + 2 \times 0.1 + 3 \times 0.1 = 1.3$. If we also know that the interval between arrivals is equally likely to be any exact number of minutes between 1 and 9, it is easy to see that the mean time between arrivals is 5 minutes. This implies we expect, on average, 1.3 persons to arrive every 5 minutes. Service is provided to one person at a time (e.g. if three arrive they are served one after the other). If service time is always exactly 4 minutes per person, the average service time per arrival is $4 \times 1.3 = 5.2$ minutes. Since this exceeds the average time of 5 minutes between arrivals, it seems likely that the queue will build up. If the service time were only three minutes per person, the average service time per arrival would be $3 \times 1.3 = 3.9$ minutes. Thus, we might expect that servers would *on average* be idle for 1.1 minutes between arrivals.

These arguments can be checked by simulation. We need pairs of random digits: the first to represent the number of people per arrival (if 0 to 7 means one person, 8 two and 9 three people, we have correct probabilities); the second to specify one of the nine possible intervals between arrivals ($1 = 1$ min, $2 = 2$ min, ..., $9 = 9$ min give correct probabilities). Since there are only nine equally likely intervals, we do not use zero as a second digit; it is easy to program a computer to 'choose' another digit if a zero occurs.

### A stable queue

Consider first a uniform service time of 3 minutes per person. According to our argument we would, in a period with thirty arrivals, expect the server to be idle for about $30 \times 1.1 = 33$ minutes. Table 10.13 sets out results for a simulation of thirty arrivals for this queueing system.

The first two columns contain random number pairs for each arrival: the first tells us how many people arrive at one time; the second (between 1 and 9) gives the time in minutes between successive arrivals. (These digits were obtained from Table 10.4 continuing pairwise from where we left off when forming Table 10.5, i.e. with pairs 71, 06, 64, etc.) In this particular example one zero second digit had to be omitted. We assume the system opens for service at 9 a.m. The first number pair (7, 1) is interpreted to mean there is a single arrival (since 0 to 7 represent a single arrival) at 9.01, one minute after the system opens. Thus, we enter 9.01 in the *arrival time* column. Since the server is unoccupied until this customer arrives, service begins im-

*Table 10.13.* Thirty arrivals simulated for a queueing system: 3 minute service time

| Random nos. | No. arriving | Arrival time | Serve start | Serve end | Free time |
|---|---|---|---|---|---|
| 7 1 | 1 | 9.01 | 9.01 | 9.04 | 1 |
| 0 6 | 1 | 9.07 | 9.07 | 9.10 | 3 |
| 6 4 | 1 | 9.11 | 9.11 | 9.14 | 1 |
| 4 8 | 1 | 9.19 | 9.19 | 9.22 | 5 |
| 4 6 | 1 | 9.25 | 9.25 | 9.28 | 3 |
| 7 9 | 1 | 9.34 | 9.34 | 9.37 | 6 |
| 0 8 | 1 | 9.42 | 9.42 | 9.45 | 5 |
| 9 2 | 3 | 9.44 | 9.45 | 9.54 | |
| 9 9 | 3 | 9.53 | 9.54 | 10.03 | |
| 2 1 | 1 | 9.54 | 10.03 | 10.06 | |
| 9 1 | 3 | 9.55 | 10.06 | 10.15 | |
| 8 2 | 2 | 9.57 | 10.15 | 10.21 | |
| 5 6 | 1 | 10.03 | 10.21 | 10.24 | |
| 6 6 | 1 | 10.09 | 10.24 | 10.27 | |
| 7 6 | 1 | 10.15 | 10.27 | 10.30 | |
| 4 5 | 1 | 10.20 | 10.30 | 10.33 | |
| 2 2 | 1 | 10.22 | 10.33 | 10.36 | |
| 7 4 | 1 | 10.26 | 10.36 | 10.39 | |
| 9 6 | 3 | 10.32 | 10.39 | 10.48 | |
| 4 4 | 1 | 10.36 | 10.48 | 10.51 | |
| 2 6 | 1 | 10.42 | 10.51 | 10.54 | |
| 3 3 | 1 | 10.45 | 10.54 | 10.57 | |
| 1 1 | 1 | 10.46 | 10.57 | 11.00 | |
| 4 5 | 1 | 10.51 | 11.00 | 11.03 | |
| 2 2 | 1 | 10.53 | 11.03 | 11.06 | |
| 0 6 | 1 | 10.59 | 101.6 | 11.09 | |
| 7 8 | 1 | 11.07 | 11.09 | 11.12 | |
| 6 4 | 1 | 11.11 | 11.12 | 11.15 | |
| 9 5 | 3 | 11.16 | 11.16 | 11.25 | 1 |
| 9 4 | 3 | 11.20 | 11.25 | 11.34 | |
| | | | | Total free time | 25 |

mediately; it takes 3 minutes and so is completed at 9.04, this being entered in the relevant column. In the column headed *free time* we enter any free time for the server prior to that customer's arrival. Here it is 1 minute (between opening time of 9.00 and first arrival time of 9.01).

The next number pair (0, 6) implies one person arrives 6 minutes after the previous arrival. Thus, the time of arrival is 9.07 (the previous arrival was at 9.01). As the server is free, service commences at once and is completed at 9.10. An entry 3 is put in the final column as the server was free from 9.04 to 9.07.

We proceed in this way. When we get to the eighth arrival, the random number pair (9, 2) implies that three people arrive 2 minutes after the previous (single) arrival. As the server is busy, they have to wait 1 minute, then the three new arrivals are served one at a time, so the total service time for *all three* is 9 minutes. Note that between the eighth and twelfth arrivals we get a run of arrivals mostly with more than one person at a time – respectively, 3, 3, 1, 3, 2. Multiple arrivals have a relatively low probability compared with single arrivals, but this sort of 'bunched' behaviour in a queue will occur from time to time (part of the 'uncertain' or 'random' nature). Arrivals after this 'rush' must wait for service, but by the time we get to the twenty-eighth arrival the server has again caught up. We may think of the system starting again from scratch at this point. Adding all entries in the free time column, we find these come to 25 minutes compared with our predicted 33 per thirty arrivals. This sort of discrepancy from expectation is not surprising in a short run. A simulation over some thousand arrivals could be expected to bring the free time expectation closer to the predicted 1.1 minute per arrival. On p. 171 we found the expected number of persons per arrival was 1.3. Adding numbers at each arrival (column 3) and dividing by 30 (the number of arrivals), we find the average numbers of persons per arrival to be $43/30 = 1.43$. Again, this discrepancy is not surprising in a relatively short run.

### A time change

If we alter the service time from 3 to 4 minutes per customer and study the queueing behaviour for the *same* thirty arrivals, we get the pattern in Table 10.14. The first four columns are identical with those in Table 10.13 since they represent the same arrivals. The remaining columns reflect the change in service time to 4

*Table 10.14.* Thirty arrivals simulated for a queueing system: 4 minute service time

| Random nos. | No. arriving | Arrival time | Serve start | Serve end | Free time |
|---|---|---|---|---|---|
| 7 1 | 1 | 9.01 | 9.01 | 9.05 | 1 |
| 0 6 | 1 | 9.07 | 9.07 | 9.11 | 2 |
| 6 4 | 1 | 9.11 | 9.11 | 9.15 | |
| 4 8 | 1 | 9.19 | 9.19 | 9.23 | 4 |
| 4 6 | 1 | 9.25 | 9.25 | 9.29 | 2 |
| 7 9 | 1 | 9.34 | 9.34 | 9.38 | 5 |
| 0 8 | 1 | 9.42 | 9.42 | 9.46 | 4 |
| 9 2 | 3 | 9.44 | 9.46 | 9.58 | |
| 9 9 | 3 | 9.53 | 9.58 | 10.10 | |
| 2 1 | 1 | 9.54 | 10.10 | 10.14 | |
| 9 1 | 3 | 9.55 | 10.14 | 10.26 | |
| 8 2 | 2 | 9.57 | 10.26 | 10.34 | |
| 5 6 | 1 | 10.03 | 10.34 | 10.38 | |
| 6 6 | 1 | 10.09 | 10.38 | 10.42 | |
| 7 6 | 1 | 10.15 | 10.42 | 10.46 | |
| 4 5 | 1 | 10.20 | 10.46 | 10.50 | |
| 2 2 | 1 | 10.22 | 10.50 | 10.54 | |
| 7 4 | 1 | 10.26 | 10.54 | 10.58 | |
| 9 6 | 3 | 10.32 | 10.58 | 11.10 | |
| 4 4 | 1 | 10.36 | 11.10 | 11.14 | |
| 2 6 | 1 | 10.42 | 11.14 | 11.18 | |
| 3 3 | 1 | 10.45 | 11.18 | 11.22 | |
| 1 1 | 1 | 10.46 | 11.22 | 11.26 | |
| 4 5 | 1 | 10.51 | 11.26 | 11.30 | |
| 2 2 | 1 | 10.53 | 11.30 | 11.34 | |
| 0 6 | 1 | 10.59 | 11.34 | 11.38 | |
| 7 8 | 1 | 11.07 | 11.38 | 11.42 | |
| 6 4 | 1 | 11.11 | 11.42 | 11.46 | |
| 9 5 | 3 | 11.16 | 11.46 | 11.58 | |
| 9 4 | 3 | 11.20 | 11.58 | 12.10 | |
| | | | | Total free time | 18 |

minutes per customer. Note that at the start the server has some free time, but once we meet that run of multiple arrivals the queue soon builds up and, because of the longer service time, it continues to do so. After about 10 a.m. the wait is seldom less than half an hour. If we continued this simulation, the waiting time would tend to get progressively longer.

What would happen in practice with a situation like this? Almost certainly there would eventually be what we call a change in *queue discipline*. Some potential arrivals, seeing a long queue, will go away; or the server might relieve the situation by working more quickly to speed up serving time to reduce customer wait; or an extra server may be brought in until queue length is reduced.

### More realistic queues

The queues just discussed are somewhat artificial, although they might be reasonable approximations in a situation where people arrive at a service point in, say, an elevator which can complete a journey and discharge passengers (if any) only at one minute intervals; then, on arrival, each has to provide particulars at a registration desk, this taking a fixed time for each person.

Queues like those in a post office are theoretically, and in practice, more complicated. People turn up (usually singly) in a more haphazard way (the rate depending on time of day) and then join a queue which may lead to several service points. The favoured system in a modern post office is for people to form one queue and the person at the head to go to the first free service point. Service times may be extremely variable: the time to purchase a first-class postage stamp is very different from that required to retax a motor-vehicle; an intermediate time will be needed to pay a pension order.

A fairly simple queueing system that has been studied in detail mathematically is one in which arrivals are single and *at random* at a specified average rate per unit time. The numbers arriving in a given time interval will then have a Poisson distribution (see p. 103). If customers are served one at a time in order of arrival and the service time has an exponential distribution (p. 107), it is not difficult to show (and not surprising) that the queue will grow indefinitely if the average time for servicing exceeds the average time between arrivals. The theory exists to work out the probabilities of various queue lengths at any given time, the proportion of time the server will be free, etc., when the average service time is less than the average time between arrivals.

Real queueing systems often resemble the above *Poisson–exponential* system at least over short time intervals. Queueing theory is important not only in studying queues of people; much of the basic theory was developed in studies of telephone traffic through exchanges. It may also be applied to patterns of transporters arriving and loading at freight depots or ships at ports.

## Practical simulations

Simulations are invaluable for studying, not only the effect of changes in a single queue (introducing extra servers, streamlining to reduce service times, introducing different service points for different requirements, etc.), but are also ideal for planning interrelated queueing systems. Design of a modern airport, where there is a complex of different queues, is a good example. Before building new major airports, mathematical models are set up that 'simulate' the various facets and the different forms of uncertainty. O.R. methods play a key role and the speed of the computer enables planners to see quickly the effect of possible design changes. Here are some of the factors that must be allowed for. Typically, when a passenger arrives to board a flight, he queues first at the check-in desks, or earlier at ticket-sales if he has not already booked. At some airports seat allocation is made at a separate point to the baggage check-in. Then there is the queue for a security search and, if the journey is international, also for emigration formalities, and sometimes also for customs clearance. At some stage, while braving this mêlée, we may queue for refreshments and also for the traditional duty-free bad bargain. Meantime, arriving passengers must be kept separate from the departure flow. International arrivals first queue for immigration, then to collect baggage, again to go through customs; there is another queue for transfer passengers to check bookings; and further queues for taxis, buses or to buy train tickets for the journey from airport to city centre. While passengers are being processed air-traffic controllers are dealing with queues of aircraft wanting to land, take off or disembark their passengers and baggage at a terminal pier. Refuellers and caterers and engineers are being scheduled to supply and service aircraft.

Simulations can be invaluable in studying how a proposed airport layout is likely to cope with various forecast traffic flows. The key to good design of an air terminal is elimination of bottlenecks. It is little use speeding up baggage delivery from aircraft if there are insufficient customs channels to clear passengers speedily. It is silly to load passengers on to a plane if it has little hope of getting a take-off clearance for several hours. On top of obvious potentials for delay, there is always the unexpected. I was one of many who suffered the chaos at London Heathrow on 6 October 1986 when the main air-traffic control computer went out of action for several hours. Yes, there is even an element of uncertainty about the mighty computer!

Not only new airports, but hospitals, large manufacturing plants and warehouses are all happy hunting grounds for simulation studies. It is much cheaper to study the effect of increasing the number of furnaces in a steel mill from two to three on demand for raw materials, output and profits with the help of a mathematical model and computer simulation than it is to build the extra furnace and see what happens, always providing the simulation model is a reasonable reflection of reality. Simulation studies are very much a part and parcel of operational research.

## Marketing and quality

Simulation is only a small but growing sector of industrial or commercial applications of statistics. We look at one other industrial application.

Mass production and marketing competition has focused attention on quality. Acceptance sampling is one *quality control* technique used to monitor the overall standard of items produced in bulk.

Economic reality usually means that not every item is up to standard. In large batches some items are bound to be defective. To eliminate all would often make the product prohibitively expensive. Goods must comply with legal requirements for weights and measures, a matter discussed in Chapter 13, but, even if legal requirements are not dominant, customer expectation may be.

Only if it is simple to inspect every item and easy to detect nearly all faulty ones and remove them will a complete inspection be made. This is usually impossible timewise or costwise. Sometimes the only way to test an item is to destroy it. If electrical components should have a minimum life of 1000 hours one cannot run a pre-marketing check on all to see how long each operates before failure.

A manufacturer offers a guarantee on a basis of what is economically sensible, including his assessment of a wise policy to maintain goodwill. He may agree with the purchaser that 1% of items defective in a large batch is tolerable; 5% is not. Having agreed acceptable quality limits with a buyer, the manufacturer (or sometimes the buyer) usually checks quality using a sample.

He may take a random sample (p. 107) of items from each large batch. Each item in the sample is tested and the number of faulty items noted. A small proportion faulty in the sample implies a high probability there will be only a small proportion faulty in the batch. In practice, a number is specified in advance and, if there are more than that number of faulty items in the sample, the whole batch is rejected. For rejected batches, if testing is non-destructive and not too expensive, every item might be checked and all faulty ones removed; otherwise, the batch may be scrapped or sold as *second grade*.

The rule for testing might be to accept a batch if and only if we find not more than two defective in a random sample of 200. Such a scheme is called an *acceptance sampling scheme*.

## Implications

What are the implications? Suppose in the large batch 1% (one item in 100) are defective. Then the probability any item in the sample is defective is approximately $p = 0.01$. Approximate, because after the first unit is selected the probability each later item is faulty is *conditional* upon the number of faulty items already selected. However, if the batch size $N$ is appreciably larger than the sample size $n$, these conditional probabilities differ little from the unconditional probability of 0.01. For example, if $N = 10\ 000$

and one hundred are faulty, Pr(first faulty) $= \frac{1}{100}$. Now Pr(second faulty | first faulty) $= \frac{99}{9999}$ and Pr(second faulty | first good) $= \frac{100}{9999}$. This follows from our definition of *conditional probability* Pr$(A | B)$ given on p. 61. Both conditional probabilities approximate the unconditional probability 0.01 to an accuracy of four decimal places.

All this means that, if there are 1% defectives in a large batch, then, in a sample of 200, the number defective in the sample has (approximately) a binomial distribution (p. 78) with $n = 200$ and $p = 0.01$. Further, we know (p. 103), if $n$ is large and $p$ small, the binomial distribution is very like a Poisson distribution with expectation $np = 200 \times 0.01 = 2$ (p. 104).

We gave the probabilities for that distribution in Table 6.7 (p. 104), from where we see that the probability of getting between zero and two defectives inclusive (by adding the probabilities of each) is 0.677. Since we accept batches with two or less defectives, this is the probability of accepting a batch *if* the true level of defectives is 1%.

### How many defectives?

So far so good, but buyer and seller will also be interested in the probability of accepting batches with other percentage defectives if we use the rule *accept only if a sample of 200 contains two or less defective*.

We may work out the probability of accepting batches with various proportions of defectives by using appropriate tables for the Poisson distribution. We omit details, but Table 10.15 shows probabilities that we find two or less defectives in the sample when there are various percentage defectives in the large batch. Obviously, when there are no defectives in a batch, the probability of acceptance is 1 because there can then be no defectives in the sample.

It is clear from the table that nearly one in three samples (a proportion of $1 - 0.667 = 0.323$) would be rejected if only 1% were defective. Also, if 2% were defective, we would still *accept* the batch for nearly one-quarter of all samples (a proportion of 0.238).

*Table 10.15.* Acceptance probability for various per cent defectives in batch:
$n = 200$, two or less defective in sample

| % defective: | 0 | 0.5 | 1 | 1.5 | 2 | 3 | 4 | 5 | 6 |
|---|---|---|---|---|---|---|---|---|---|
| Prob. acc.: | 1 | 0.920 | 0.677 | 0.423 | 0.238 | 0.062 | 0.014 | 0.003 | 0.001 |

*Table 10.16.* Acceptance probability for various per cent defectives in batch:
$n = 400$, four or less defective in sample

| % defective: | 0 | 0.5 | 1 | 1.5 | 2 | 2.5 | 3 |
|---|---|---|---|---|---|---|---|
| Prob. acc.: | 1 | 0.947 | 0.629 | 0.285 | 0.100 | 0.029 | 0.008 |

We note too that the probability of accepting batches decreases as the per cent defectives in the batches increases. Intuitively, this is what we would expect.

A buyer might, however, not be very happy with the above scheme, for it would indicate over 6% (a proportion of 0.062) of batches with 3% defective would be accepted.

There is no theoretical restriction on choice of sample size or in setting a maximum number of faulty items for which we accept a batch. Tables exist that give sample schemes with characteristics acceptable to both producer and consumer, ensuring a small probability that good batches (those with few defectives) will be rejected (something the producer does not want) and also a small probability that batches with many defectives will be accepted (something the consumer does not want).

We may reduce the probability of accepting batches with many defectives either by increasing sample size or reducing the maximum number (proportion) of defectives allowing acceptance, or both.

Table 10.16 gives the probabilities of accepting a batch with various per cent defectives if we take a sample of 400 and accept only if there are not more than four defectives.

Clearly, with this scheme, we are unlikely to accept many batches with more than 2.5% defective, and we have only reduced slightly (from 0.677 to 0.629) the probability of accepting a batch with 1% defective. We have increased (compared with Table

10.15) the probability of accepting a batch with 0.5% defective. Many elaborations of acceptance sampling schemes are possible.

Quality control methods (of which acceptance sampling is just one) are an expanding area in statistics. Japanese manufacturers owe a lot of their success in world markets to the rigour with which they employ these methods, originally developed during and after World War II in the U.K. and the U.S.A. There is a growing awareness of the need to apply these methods in their countries of origin, something manufacturers have been very slack about until recently.

### A review

The situation in this chapter contrasts strongly with the one we met when studying the environment. The uncertainties we have examined here – by simulation and an example of quality control – are cold and clinical; the emotional content is virtually nil.

In an industrial context classical probability theory and statistics provide the background for experiments to improve efficiency. We have illustrated the ideas here by over-simplified examples, but these demonstrate the principles.

One wonders whether the benefits that these approaches bring to industry and their cold clinical nature may not have helped harden industrialists' attitudes to more humanistic uncertainties and risks. Has success of these formal applications fuelled the reluctance of some captains of industry to face up to more emotive issues like dangers to health and the environment and left them realizing that appropriate action to remove such risks may negate the higher cost efficiency brought to them by operational research and statistics?

# HAZARDS: MOVING AND STATIONARY

## Not much safety on the road

In this chapter we look at situations where conclusions about risks come largely from common-sense interpretations of available data. Sometimes a further analysis using classical statistical methods may result in subtle modifications of what at first seem obvious conclusions from an intelligent inspection of data. The interpretation of road accident statistics in relation to wearing of seat belts is a good example.

From 1980 to 1985 between five and six thousand people died in road accidents in the U.K. each year; over 70 000 were seriously injured.

Evidence has mounted that wearing seat belts reduces road casualties. Police estimate compliance with U.K. regulations requiring seat belts to be worn by drivers and front-seat passengers at over 90%. Table 11.1 shows averages for the three years prior to, and the first three years after, the legislation was introduced. We give averages for *all road casualties*, for those *seriously injured* and those *killed*. Means are rounded to the nearest 1000 for injuries and the nearest ten for deaths. This is sensible rounding because death is a clearly defined category and data thereon reasonably accurate, whereas statistics on injury are subject to availability and are often under-reported. There is also subjectivity in defining *serious injury*.

The drop in all averages is *prima facie* evidence of a beneficial effect of seat belt legislation, but in Chapter 3 we pointed out potential pitfalls in taking totals or averages at face value. Might

Table *11.1*. U.K. road casualties, three year averages

|  | Total | Seriously injured | Killed |
|---|---|---|---|
| 1980–82: | 329 000 | 79 000 | 5930 |
| 1983–5: | 318 000 | 72 000 | 5410 |
| Fall in average: | 11 000 | 7 000 | 520 |

there be a long-term downward trend irrespective of legislation? Could another factor (weather, road improvements, less travelling, etc.) explain the fall? Could a drop for front-seat drivers and passengers be partially offset by higher casualties among other road users? This might happen if the extra security for motorists led to careless driving habits that put cyclists and pedestrians at increased risk.

### A detailed analysis

To study these possibilities an elaborate analysis was undertaken on behalf of the U.K. Department of Transport by James Durbin and A. C. Harvey of the London School of Economics. They used sophisticated statistical methods, the details being given in the *Journal of the Royal Statistical Society* (Harvey and Durbin, 1986).

For casualties broken down by categories of road users, they found a real (statistically highly significant, p. 100) drop in drivers and front-seat passengers killed and seriously injured after belts were made compulsory. There was evidence too of a small increase in cyclists and pedestrians *killed*, but little change for these groups in totals *killed and seriously injured*. They reached these conclusions after allowing for any trends that might be independent of seat-belt law. The small increases in cyclist and pedestrian deaths could reflect greater driver carelessness, but, if so, why was it offset by a compensating decrease in numbers seriously injured? We have pointed out that figures for serious injury are often incomplete and there is subjectivity in defining *serious injury*, but there is no obvious reason why reporting should be worse (or better) after seat belts were made compulsory. We cannot ignore unexplained fluctuations, but these small ones do not annul stronger indications of overall benefit from seat belts.

A surprise result from Durbin and Harvey's analysis was a significant increase in rear-seat deaths (but again not in deaths plus serious injury). It has been suggested this might be due to a change in seating habits – more people travelling in rear seats to avoid having to belt up. A novel idea put forward by eminent statistician Violet Cane was that in accidents rear-seat passengers are now subject to a harder impact with those in front, since the latter now provide a less flexible barrier because they are constrained by seat belts. Another possibility is that growing use of headrests to reduce the danger of whiplash injury to front-seat occupants provides an additional hard barrier if rear-seat passengers are thrown forward.

A detailed and critical discussion is published with the paper by Harvey and Durbin. The authors' rebuttal of most criticisms of their findings is convincing, but continuing research is needed on some points.

### Presenting statistics

Accident (and many other) statistics can be presented in a variety of forms. Numbers killed or injured on roads depend upon numbers of road users, the total mileage covered by all vehicles, the numbers of passengers per vehicle and the degree of pedestrian exposure. Alterations in average or allowable speeds, improvements in vehicle safety and elimination of traffic hazards are also relevant. Harvey and Durbin took such factors into account so far as possible.

Accurate figures for total road usage are clearly hard to obtain. Estimates of total vehicle mileage may be based on numbers of registered vehicles and reasonable assumptions about the average distance travelled by different types. Sampling methods may provide these estimates. U.K. statistics indicate that for *drivers only* the number killed per hundred million vehicle kilometres has dropped from about 0.9 in 1975 to about 0.5 in 1985.

We might also consider casualties per 100 000 population. In the U.K. the numbers of road deaths (all users) per 100 000 population in 1985 was about 9.1; almost exactly one-half of the corresponding figure for the U.S.A. Yet, on a vehicle mileage basis, U.S. figures for road deaths are marginally lower than those for the U.K. This is because the U.S.A. has more cars per head of population and average annual mileages may be greater, although reliable figures for the latter are difficult to obtain.

Problems of finding the most appropriate base for comparisons are not confined to transport: we met other examples in Chapter 2. Pearce (1987) illustrates this problem for sulphur emissions, that problem we looked at in Chapter 8. For total sulphur emissions in Europe, the U.S.S.R leads with the U.K., East Germany and Italy close behind. On emissions per head of population, East Germany wins easily, followed by Czechoslovakia and Hungary; the U.K. is now eleventh. Based on emission per square kilometre of land, East Germany is again top with the U.K., Czechoslovakia and the Netherlands tightly grouped as runners up, but well behind the leader.

### Safety in the air

In 1985, a bad year for aviation, 2129 people were killed in scheduled and non-scheduled commercial flying accidents worldwide. World statistics for air accidents involving company aircraft, private business and pleasure flights, air-taxi work, and so on – known as the *general aviation sector* – are hard to obtain, but indications are that there were probably some three thousand deaths in 1985 in this sector worldwide. We look at data from the U.S.A (by far the largest general aviation users) on p. 191.

Table 11.2 gives the total commercial passenger flying fatalities (whole world) for each of the years 1980–85.

For comparison, road death statistics for the U.K. and U.S.A. for 1980–85 are given in Table 11.3. Note the high year-to-year variability for air compared to road deaths in any one country. The greater variation in Table 11.2 reflects the varied size of incidents. The worst air accident in 1985 took place on 12 August when a Japanese airliner crashed into Mt Ogura killing 520 people after decompression damaged a fin and resulted in control failure.

The Japanese crash was unusual in nature, but any large

Table 11.2. Air passenger deaths,
scheduled and charter flights

| Year | Number killed |
|------|---------------|
| 1980 | 1329 |
| 1981 | 710 |
| 1982 | 1012 |
| 1983 | 1202 |
| 1984 | 451 |
| 1985 | 2129 |

Source: C.A.A. World Airline Accidents
Summary, 1986.

Table 11.3. Road deaths, 1980–85

| Year | U.K. | U.S.A. |
|------|------|--------|
| 1980 | 5953 | 51 100 |
| 1981 | 5846 | 49 300 |
| 1982 | 5934 | 43 900 |
| 1983 | 5445 | 42 600 |
| 1984 | 5599 | 44 200 |
| 1985 | 5165 | 43 000 |

Sources: C.S.O. Annual Abstracts of Statistics, 1987, HMSO. U.S. Bureau of Census: Statistical Abstracts of the United States, 1986, Washington, D.C.

accident, or risk of one with a high likelihood of many fatalities, causes greater public apprehension than more common accidents resulting only in isolated deaths, a point we noted about attitudes towards risks in Chapter 2, p. 34.

World figures for commercial flying accidents disguise national differences. The magazine *Flight International* publishes each year comprehensive statistics on air-transport accidents. Over many years, Australia has a consistently good safety record. At the other extreme, Colombia has a consistently dismal record. Such differences may reflect, not only variations in pilot and maintenance standards, in aircraft age and type, but also flying conditions (e.g. weather and the terrain over which most flights are made).

Australian weather conditions are generally favourable for flying, the navigational aids adequate and the terrain in general not mountainous. Many Colombian flights cross the Andes. Britain and the U.S.A. usually emerge well from these safety reviews, but concern has been expressed that U.S. safety standards may be succumbing to commercial pressures following the controversial 'deregulation' of U.S. air transport. The statistical evidence is not yet sufficient to confirm a slippage; we are looking at small changes in small probabilities (see p. 92).

### Fly or drive?

To compare road and air casualties, we need some rational common basis.

Most of us spend less time in aeroplanes than motor-cars. Possible bases for safety comparisons would be total passenger hours spent in planes and in cars each year, or corresponding figures for passenger or vehicle miles rather than hours. It is hard to get accurate figures for road vehicles on a time basis, but we indicated (p. 185) that, with reasonable assumptions about annual mileages, the U.K. death rate for *drivers only* is currently about 0.5 per hundred million vehicle kilometres.

The more stringent requirements to keep air-transport records enables us to get reasonable figures for passenger hours or passenger miles (or kilometres) for air travel. Airline casualties are often based on fatalities per unit time (100 000 hours) or per 100 000 km flown, or per million passenger kilometres flown.

The apparent magnitude of any change in accident rates depends on choice of base. Tables 11.4 and 11.5 give figures for various years for scheduled fixed wing passenger services by U.K. airlines.

With each of the three bases in Table 11.4, there is an indication of improving safety, but each base shows a different rate of change with time. In recent years route lengths have tended to increase with greater distances between refuelling stops (especially on international services), so stages tend to be longer. Larger aircraft now carry more passengers. You might like to think of any implica-

*Table 11.4.* Some measures of changing U.K. air safety in several periods

| Years | Thousands of aircraft flight stages per fatal accident | Millions of aircraft km per fatal accident | Millions of passenger km per passenger killed |
|---|---|---|---|
| 1950–54 | 107.4 | 61.8 | 50.0 |
| 1970–74 | 897.4 | 737.6 | 657.7 |
| 1975–79 | 1797.2 | 1481.6 | 3240.0 |
| 1980–84* | — | — | — |

*There were no accidental deaths in this period. Four passengers were seriously injured.
*Source:* C.S.O. Annual Abstracts of Statistics, 1987, HMSO.

*Table 11.5.* Alternative measures of U.K. air safety

| Years | Fatal accidents per 100 000 aircraft flight stages | Fatal accidents per hundred million aircraft km | Passengers killed per hundred million passenger km |
|---|---|---|---|
| 1950–54 | 0.93 | 1.62 | 1.99 |
| 1970–74 | 0.11 | 0.14 | 0.15 |
| 1975–79 | 0.06 | 0.07 | 0.03 |
| 1980–84 | 0.00 | 0.00 | 0.00 |

tions about the safety of such aircraft by comparing figures in the final two columns of Table 11.4 for each of the periods. The period 1980–84 was an extremely happy one for British aviation. Unfortunately, it was already clear by the end of 1986 that the 1985–89 return will not be a row of blank entries.

Table 11.5 is derived from Table 11.4 (with some rounding).

These tables indicate air travel per unit distance is safer than road travel. Since air travellers cover appreciable distances in a relatively short time, the situation may be very different if looked at on a time-exposure basis. On a flight from the U.K. to Australia, a passenger covers in twenty-four hours about the same or a greater distance than most people travel by car in a year.

*Table* 11.6. U.S. scheduled air services: accident statistics

| Year: | 1979 | 1980 | 1981 | 1982 | 1983 | 1984 |
|---|---|---|---|---|---|---|
| Total no. of accidents: | 18 | 19 | 25 | 16 | 21 | 12 |
| No. of fatal accidents: | 3 | 2 | 4 | 5 | 4 | 1 |
| Numbers killed: | 352 | 14 | 4 | 235 | 15 | 4 |

*Source:* U.S. Bureau of the Census. Statistical Abstracts of the United States, 1987, Washington, D.C.

## General aviation

The safety of scheduled and charter flights is, generally speaking, better than that in the general aviation sector (p. 186) of the aviation industry.

To give a basis for comparison, Table 11.6 presents data for scheduled U.S. domestic and international carriers. Once again, note the large variation from year to year in the number of fatalities. Until 1982 the passenger miles flown was about 250 billion (thousand million) per annum; it rose to about 280 billion in 1983 and 300 billion in 1984, with no evidence of an increase in accidents.

For general aviation in the U.S.A., involving large numbers of business and private light aircraft, flown by pilots with varied qualifications and experience, there is greater annual uniformity both in the numbers of fatal accidents and in totals killed. This is because large numbers of smaller aircraft are involved and the numbers of deaths *per incident* tends to be small, while the number of incidents (compared to scheduled and charter flights) is large. The number of fatal accidents per 100 000 aircraft hours is very much higher than for scheduled carriers. Some details are given in Table 11.7.

To give a cross-comparison basis, the numbers killed each year in general aviation accidents in the U.S.A. is about 20% of the numbers killed on roads in the U.K., or slightly more than the number of motor-cyclists and pillion riders killed each year in the U.K. This is clear from figures in Table 11.8, which gives a break-down of road casualties in the U.K. over an eleven year period. You might like to study this table in its own right. Do you see any

*Table 11.7.* United States general aviation: fatal accident data

| Year: | 1979 | 1980 | 1981 | 1982 | 1983 | 1984 |
|---|---|---|---|---|---|---|
| No. of fatal accidents: | 638 | 622 | 654 | 589 | 547 | 529 |
| Fatal accident/ 100 000 ac/h: | 1.65 | 1.71 | 1.84 | 1.84 | 1.76 | 1.73 |
| Numbers killed: | 1237 | 1252 | 1282 | 1182 | 1046 | 998 |

*Source:* U.S. Bureau of the Census. Statistical Abstracts of the United States, 1986, Washington, D.C.

*Table 11.8.* Breakdown of deaths on U.K. roads, 1975–85

| Year | Total killed | Pedestrians | Pedal cyclists | Two-wheel vehicles | Cars, taxis | Other |
|---|---|---|---|---|---|---|
| 1975 | 6366 | 2344 | 278 | 838 | 2444 | 462 |
| 1976 | 6570 | 2335 | 300 | 990 | 2520 | 425 |
| 1977 | 6614 | 2313 | 301 | 1182 | 2441 | 377 |
| 1978 | 6831 | 2427 | 316 | 1163 | 2569 | 356 |
| 1979 | 6352 | 2118 | 320 | 1160 | 2429 | 325 |
| 1980 | 5953 | 1941 | 302 | 1163 | 2278 | 269 |
| 1981 | 5846 | 1874 | 310 | 1131 | 2287 | 244 |
| 1982 | 5934 | 1869 | 294 | 1090 | 2443 | 238 |
| 1983 | 5445 | 1914 | 323 | 963 | 2019 | 226 |
| 1984 | 5599 | 1868 | 345 | 967 | 2179 | 240 |
| 1985 | 5165 | 1789 | 286 | 796 | 2061 | 233 |

*Source:* C.S.O. Annual Abstract of Statistics, 1987, HMSO.

obvious evidence from it that compulsory wearing of seat belts (from 1983 onward) has led to increasing death rates in any group of road users? Who seems to benefit? You may feel your conclusions are contrary to some of those reached by Harvey and Durbin that we mentioned on p. 184, but remember they allowed for long-term trends in accident statistics not associated with changes in seat-belt legislation.

Breakdowns of air accident statistics are also informative. For example, detailed figures for U.S. general aviation accidents in 1983 published in the magazine *Flight International* indicated that the death rate per 100 000 hours flown was nearly seven times as great for single piston-engined fixed-wing aircraft as it was for

Table 11.9. Rail fatalities in the U.K., 1975–1985

| Year | Deaths in train accidents | Deaths in other rail related accidents | Other deaths on railway premises* |
|------|---------------------------|----------------------------------------|-----------------------------------|
| 1975 | 57 | 58 | 12 |
| 1976 | 18 | 65 | 14 |
| 1977 | 12 | 61 | 11 |
| 1978 | 22 | 76 | 16 |
| 1979 | 20 | 74 | 13 |
| 1980 | 7 | 52 | 10 |
| 1981 | 7 | 58 | 11 |
| 1982 | 11 | 40 | 6 |
| 1983 | 10 | 53 | 6 |
| 1984 | 30 | 41 | 5 |
| 1985 | 6 | 55 | 13 |

*Excluding suicides or accidents attributable directly to unlawful trespass on railway property.

Source: C.S.O. Annual Abstracts of Statistics, 1987, HMSO.

turbojets. It was nearly twice as high for helicopters as for single piston-engined fixed-wing aircraft.

The accident rate for general aviation aircraft being used for business and pleasure was over six times as high as that for accidents during pilot training. Do some people not learn by experience?

Like the commercial aviation sector, general aviation has its 'bad' years. In the U.K. there were more fatal accidents in this sector in 1987 than for many years, and preliminary findings indicate pilot error was a common cause.

## On the trains

We suggested in Chapter 1 that, if rail casualties were as high as those on U.K. roads, there would be a public outcry. Table 11.9 gives some U.K. rail accident data.

The first column shows the most variability. This is because a major train accident is likely to involve multiple deaths. The situation is less extreme than that in flying, but the same principles apply. The second column includes deaths by accident to rail travellers or staff where an individual rather than the whole train is involved. This includes death by falling from a train, accidents when alighting or boarding, and so on. Other accidents include those in railway yards or associated with railway operations, but not involving trains.

To give some idea of the relative safety in each of the above years, the total passenger kilometres involved was about thirty billion.

### Safe journey

Now we have data for U.K. road, rail and air casualties we can produce a reasonable answer to the question 'What is the safest way to travel from Glasgow to London?' We quoted a driver death rate of approximately 0.5 per hundred million driver kilometres on p. 185. There is a greater year-to-year variation for corresponding figures for rail or air passengers, but over recent years in the U.K. neither figure has exceeded 0.03 deaths per hundred million passenger kilometres. From this, it is not unreasonable to conclude rail or air travel is sixteen or more times safer than driving one's own car from Glasgow to London. But note we are comparing U.K.-wide figures for car drivers with those for fare paying passengers by air or rail, and inferring they hold for a particular route. Might not the picture be different if we concentrated on data specific to that route? Is the Glasgow–London route more (or less) dangerous than the average for all U.K. roads? Are there special rail or air safety factors? Also, if we travel by road, we do not have to drive ourselves. The back-seat passenger death rate per unit distance travelled is less than that for drivers. I have no accurate figure for this difference, but on reasonable assumptions the rate is probably still appreciably higher than air or rail accident rates.

It is interesting to note that in Table 2.1 air travel appeared appreciably less safe than rail travel. However, the figures there were for 1972–76. In the interim there has been a reduction in air fatalities with little change in rail safety. This is implicit in Table 11.5 so far as air travel is concerned.

We have not mentioned serious injury; this is most common in road accidents, rarer with rail and rarer still with air accidents, where the outcome of major disasters is usually death. On the whole, we may reasonably conclude travel by rail or air is safer than road for our journey, but we cannot be too assertive about exactly how much safer. Clearly, individuals have, in one sense at least, more control over their own destiny when driving a car in that they can determine the degree of care they take. On the the other hand, they are more exposed to the eccentricities of other motorists, whereas rail and air transport operates in a more restricted environment, where train drivers and pilots and others involved with safety follow much stricter codes than those applied to the private motorist.

### Sea losses

In the fishing industry each year boats and lives are lost: commercial sea fishing is widely regarded as hazardous. To some extent, as with road accidents, we tend to accept these losses as unfortunate, paying particular attention only to a few major disasters which disrupt life in often small and closely knit fishing communities.

Major shipping disasters are comparatively rare, but the more unexpected receive a lot of publicity, a classic example being the sinking of the *Titanic* and a recent one the capsizing at Zeebrugge in 1987 of the ferry *Herald of Free Enterprise*.

As we move from road, to rail, sea or air the scale of individual accidents, or potential accidents (but not their frequency), tends to increase. Increasing potential for single major accidents has resulted in more stringent safety requirements to reduce that potential.

In the case of rail, air or sea accidents, there are rigorous inquiry procedures following all major (and even many minor) accidents. These inquiries often come up with positive recommendations, sometimes followed by appropriate legislation or action for increased safety in the future.

## The build-up to trouble

Many major accidents, whether mechanical failure or human error the key, have no single cause, but are due to an unpredicted combination of interacting factors.

Some twenty years ago a Trident aircraft took off from Heathrow Airport, London. On the climb out, an instrument called a *stall warner* gave an alarm. The crew ignored this indication that they were climbing too steeply and too slowly, for on that particular aircraft the warning device was like Hilaire Belloc's Matilda, a girl who delighted in initiating false alarms. Only this time it was for real. There are other symptoms that confirm an aircraft is really about to stall and literally fall out of the sky; pilots recognize these.

On that particular flight, the chief pilot failed to react to these other symptoms – he was the victim of a heart attack. While we shall never be sure what went on in the minds of the rest of the flight crew at the critical moment, it seems the co-pilot did not realize the danger until too late; perhaps concentration was low because crew members were alleged to have quarrelled just before take-off about an industrial dispute.

The lessons from this incident were as follows:

(i) Seek more reliable stall warners that do not give too many false alarms.
(ii) Consider the need for more rigorous medical checks (these were already strict) to detect potential serious cardiac problems.
(iii) Alert second officers to be more vigilant and ready to take control if they suspect the captain is suddenly incapacitated.

Human nature is sometimes beyond our control, but many airlines make every attempt in allotting crew duties to avoid banding together people known to experience mutual antagonism.

### Survivable?

In August 1985, fifty-three people died at Manchester airport in what looked at first sight to be a survivable incident. There was no crash impact and no interior damage to escape facilities. The loss of life was due to a succession of mishaps and errors. As the plane was taxiing prior to take off, an engine component ruptured and burst through a wing panel allowing fuel to spray on to the hot engine to cause an explosion and fire. The aircraft captain heard the explosion but thought it was a burst tyre and quickly turned off the runway as appropriate for such an incident. The crosswind drove smoke into the plane's cabin, seating material caught alight and gave off toxic fumes and an exit door jammed. Firemen were quickly on the scene but found three appliances empty of foam or water, though an inquest found this did not result in loss of life.

It was not known before the accident that a crosswind could cause the difficulty it did when a plane was on fire. The danger of toxic fumes from seating materials had been underestimated. The jammed exit door resulted from too hasty an attempt to deploy an escape chute (a natural reaction by a well-trained crew in the face of fire). There had been warnings of the type of engine fault that triggered the series of events, but it was believed by the airline that the risk was only to engines that had not had a modification applied to those on this aircraft. The dry hydrants might have been more serious in some other accident. The lessons from this incident have been taken on board by all concerned, but a proposal that airlines should provide 'smoke-hoods' for those caught in a cabin fire is still disputed; these may reduce danger of asphyxia, but hinder escape before one is incinerated.

### Fixed hazards

Not only in transport do we learn from sometimes all too bitter experience. The history of engineering and technological advances abounds in incidents where engineers and plant designers have learnt the hard way. There is a long catalogue of disasters to bridges. We recorded on p. 171 that the Tay bridge (1879) collapsed primarily because it was too 'rigid' to stand up to gales, although there were other contributory causes. A trend in reducing rigidity came full circle when the Tacoma Narrows bridge (1940) failed because it was too flexible.

Bridges involving novel designs have collapsed during construction. Engineers have learnt valuable lessons for future bridges of similar design or have concluded the basic design is unsuitable at least in that context (perhaps due to wind stresses, effects of low temperature or unsuitable foundations).

### An interesting comparison

A living monument to development in bridge design is provided by the adjacent Forth rail and road bridges near Edinburgh. The former is, in the eyes of the modern engineer, 'overdesigned'. It was built shortly after the Tay bridge disaster, where contributory faults included skimping on materials and shoddy construction (quality control was virtually non-existent in those days), as well as that basic design error of making it too rigid. A sturdy, yet not too rigid, heavily reinforced, cantilever design was used for the Forth rail bridge. The road bridge, built some eighty years later, is a modern suspension bridge and beside the rail crossing is thought by some to look frail. However, there have been no serious problems with it over more than two decades and it incorporates many lessons learnt from other suspension bridges, including the ill-fated Tacoma Narrows bridge.

Engineers say they 'need' failures to learn how to build better structures; there is an element of truth in this modern application of that old saw *experience is the best teacher*.

### Industrial dangers

In recent years a lot of attention has focused on industrial hazards – chemical plants, nuclear power stations and mining operations – and the danger of moving materials between such plants.

Some hazards, like fires or explosions in underground coal mines, are inherent in the nature of the operation. They can be reduced by various safety measures, such as, for example, constant testing for gas build-ups.

Hazards in modern chemical plants of previously untried design are less predictable. A leaky valve may allow a gas to escape. That in itself may do little harm unless it comes into contact with a liquid seeping from a fractured pipe; even then there may be no danger unless the temperature exceeds 30°C for an extended period. Flixborough and Bhopal are chemical accidents that many people say should not have happened. That they happened – and that others like them may reasonably be expected in the future – is the reality. We cannot say when or where the next similar incident may take place. What is important is that we learn from these happenings so as to reduce the likelihood of similar tragedies in future – or reduce the impact of such a tragedy if it does occur – by appropriate siting of more hazardous plants and by better inspection of dangerous equipment or equipment likely to fail.

The public is now, with some justification, often apprehensive of high-tech plants. We have already elaborated on reasons for their sometimes justified, at others apparently irrational, attitude in Chapter 2.

## What help probability and statistics?

In contrast to the problems we discussed in the previous chapter, objective probability had a more limited role in the situations in this chapter. It is important that we have as much information as possible on accident rates involving transport and industrial processes so that we can use statistically sound methods to detect trends, especially any increase in risk. However, as individuals, we are more concerned with one-off risks. Objectively, we may work out the probability of being killed in an air accident per hundred million kilometres travelled. The fact that this is small is some consolation if we only intend to fly 700 km, but the fact that among all flights the likelihood of our plane crashing is effectively infinitesimally small may still be little comfort to a nervous passenger who knows that in a given year one or two fatal air accidents are almost certain to occur. Could his flight be one of these 'one-off' events of low probability is the question that will play a prominent role in deciding his utility for flying.

## A review

In this chapter we first examined travel uncertainty and compared risks largely by looking at data in a common-sense way. This differs from the situation in the previous chapters in Part II in that here we have readily available data.

In the Salk vaccine study we had to start from scratch to collect the appropriate data, taking care we got what was relevant. Once we had that data, it was relatively simple to draw the correct conclusions. In studies of environmental pollution, our difficulty was to get data about output of pollutants and damage, and also to understand the chemistry of what was happening to primary pollutants, as well as the role played by meteorological and other factors.

When we looked at application of statistical methods to industry, we either collected real data (as in acceptance sampling) or generated 'make believe' data (as in simulation). In either case we interpreted that data using classical statistical techniques.

In contrast to transport risks, engineering risks and industrial accidents tend to be one-off affairs; it is important that we learn the lessons of each.

Chapters 7 to 11 show clearly that how we interpret matters of uncertainty depends very much on the context.

# CHAPTER 12

# DISEASES, DIAGNOSES AND DOUBTS

## Some of the problems

Uncertainty in all its guises from indeterminate risks to that associated with subjective and objective assessments of probabilities is widespread in medicine. We know the Salk vaccine was cleared as safe and effective only after extensive experiments. Deciding if a new treatment is beneficial, or which of several treatments is the most effective, is but one aspect of medical uncertainty. Others include the following:

1. *Diagnosis.* Several diseases have similar symptoms. How should one combine results of tests on blood or urine samples with clinical data to decide from which disease a patient suffers?
2. *Causes of disease.* Europeans smoked for some 200 years (American Indians much longer) before there was a hint this might cause lung cancer. Now, a century after the first suspicions, there is near certainty that smoking is *one* cause of this and some other forms of ill-health.
3. *Genetics or the environment.* Disease may be hereditary or may be due to exposure to environmental hazards, sometimes a combination. How do we decide whether diseases are hereditary? Which environmental factors are important?
   *Screening techniques.* Early detection of breast or cervical cancer (before clinical symptoms appear) may save life or avoid disfiguring surgery. Yet screening may produce false positives or miss cases. How do we determine optimal intervals between repeat screenings?

5. *Drug safety*. The thalidomide incident focused attention on dangerous side effects from new drugs. Minor side effects are common and generally acceptable if a drug offers proven therapeutic benefit. Rare, or delayed, dangerous side effects may only be detected after long usage. This is why it is important to monitor carefully the use of all new drugs.

## Diagnosis under uncertainty

Symptoms of mild poliomyelitis (p. 114) resemble those of influenza. The polio virus may be identified by laboratory tests, but, even if a virus or bacillus cannot be identified, scientific tests may still aid diagnoses when clinical symptoms of several diseases are similar. Computer-based diagnosis uses statistical methods to analyse information about a patient. Basically, a form of what statisticians call *discriminatory or discriminant analysis* is applied to relevant clinical and laboratory data with the aim of pinpointing the disease a patient is most likely to have. The computer is programmed to simulate the way an experienced doctor might approach the problem, to mirror that expert's *thought processes*. By doing so, it relieves the routine workload of doctors and may even speed up diagnoses, particularly in developing countries where medical resources are often severely stretched.

### A three-way choice

Here is a simple illustration of one way computer diagnosis works. Clinical symptoms like temperature, pulse rate and giddiness exhibited by a patient may be consistent with one of three diseases: we label these A, B and C. If we also have laboratory evaluations in appropriate units (percentages, parts per thousand, counts, etc.) for three blood components, $L_1$, $L_2$ and $L_3$, how might these help?

We first ask, 'Do the levels of $L_1$, $L_2$ and $L_3$ show real differences between diseases?' If they do, comparing levels in our current patient with those for earlier patients *correctly* diagnosed as having one of the diseases A, B or C is an obvious first step.

Because of natural biological variation, these constituents will

*Table* 12.1. Parts per thousand of three blood chemicals for patients with known diseases A, B or C

| Disease/ patient | Component level | | |
|---|---|---|---|
| | $L_1$ | $L_2$ | $L_3$ |
| A1 | 16 | 93 | 22 |
| A2 | 18 | 81 | 29 |
| A3 | 23 | 62 | 41 |
| A4 | 15 | 83 | 17 |
| A5 | 19 | 74 | 24 |
| A6 | 17 | 79 | 31 |
| B1 | 24 | 72 | 43 |
| B2 | 28 | 76 | 39 |
| B3 | 19 | 42 | 58 |
| B4 | 23 | 51 | 51 |
| B5 | 25 | 43 | 49 |
| B6 | 31 | 57 | 53 |
| C1 | 35 | 49 | 48 |
| C2 | 29 | 31 | 51 |
| C3 | 41 | 22 | 62 |
| C4 | 34 | 17 | 53 |
| C5 | 38 | 25 | 59 |
| C6 | 44 | 29 | 67 |
| X | 28 | 42 | 53 |

have different values for every patient, differences that are superimposed on any due to the disease.

Table 12.1 shows the levels of each of these constituents for three groups each of six patients (labelled 1 to 6). Patients in one group suffered from disease A, in another from B, in the third from C. We also show the levels of $L_1$, $L_2$ and $L_3$ for a new patient X. What is our 'best' diagnosis for X?

Looking at Table 12.1, we see that, for both $L_1$ and $L_3$, the values are generally higher for B than for A, and higher still for C, but there is some overlap between A and B and between B and C. There is no overlap, however, between values for the six patients with C and the six with A.

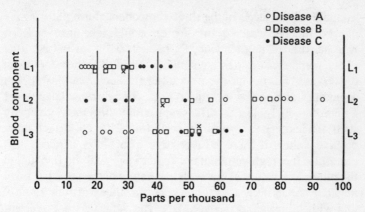

*Figure 12.1.* Levels of three blood chemicals.

Contrariwise, for $L_2$, values decrease as we move from A to B to C with overlap between A and B, and also between B and C, but again not between A and C.

We see too that, for each illness, there is more variability between patients' $L_2$ values than between their $L_1$ values. For each disease, variability among $L_3$ values is intermediate between those for $L_1$ and $L_2$.

A simple graph (Figure 12.1) makes these points clearer. In the figure, the crosses ( × ) represent the observed values for the yet undiagnosed patient X. For each constituent, his level lies outside the observed range for condition A. For $L_1$, his level is just outside the range for condition C and towards the top of the range for condition B. For $L_2$, it is near the top for condition C and equal to the lowest value for condition B. For $L_3$, his level, 53, is within the range consistent with either B or C.

This evidence *alone* tells us the new patient is unlikely to have condition A; possibly he has either B or C with the odds marginally favouring B because for $L_1$ he is just (but only just) outside the range for other patients with C.

At this stage, an experienced doctor (and a well-programmed computer) would take other factors into account. Is B common and C rare? If so, this might strengthen a belief the patient has B.

In addition, the levels of the three components for a patient may not be independent, so the doctor would take into account whether there is any tendency for patients with, say, a high level of one component also to have high levels of the others for one of the diseases, but not for the other ailments. Such relationships are sometimes obvious by looking at data, though none is apparent for the data in Table 12.1. The simplest statistical measure of any such tendency is a *correlation coefficient*. This coefficient takes positive values if there is a positive association between two quantities (i.e. high values of one tend to pair with high values of the other; low values of one with low values of the other). It takes negative values if the association goes the other way (low value of one with high values of the other). The value of the coefficient always lies between $+1$ and $-1$. Values near $+1$ or $-1$ indicate strong association; values near zero suggest it is weak (or perhaps very complicated). Statistical diagnoses using a computer and based on many correctly diagnosed cases take such relationships into account.

What about risks attached to a wrong diagnosis? If condition B responds quickly to a simple treatment (which does no harm if C is the real disease), the doctor may try that first for our patient X, looking meantime for other symptoms or doing further tests. Then, if there is no response to treatment, or new evidence points to C, he can switch to an appropriate treatment. He would not follow this line (we hope!) if the treatment for B might prove fatal if C is the true disease!

Appropriately formalized, and developed to take many more factors into account, this kind of approach is a basis for computer diagnosis. By building in various statistical notions, the computer may also estimate the probability of a wrong classification. As ever, validity depends upon getting the model right. The more past cases with a correct diagnosis we have, the better the precision of diagnosis. An amusing 'game' in which one is invited to try one's diagnostic skills was devised by Aitchison and Kay (1973).

## Gut reactions

Our example was superficial, designed only to show fundamentals. Doctors need have no fear that the computer will remove the need for their diagnostic skills, but it may save drudgery in routine situations and give backing to their gut instinct in more serious problems. The real 'doctor's dilemma' in diagnosis comes when a decision is difficult and the cost of a wrong one high. This problem often faces a surgeon in a relatively small hospital. A patient may be admitted in great pain, this due either to a relatively minor disorder that would respond in a few days to simple treatment or else symptomatic of a condition requiring an early and serious operation. Which is the patient's true condition could only be determined either by major surgery or by sending him to a larger hospital many miles away where complex scanning equipment could make a correct diagnosis. Should the surgeon operate, perhaps unnecessarily, or send the patient on a long and uncomfortable journey, meaning a delay that could be dangerous if the condition were the serious one? In such situations a good surgeon weighs imponderables in a way that relies heavily on *instinct*, combining the experience of years in a way even he would be hard put to describe, but one almost certainly requiring the weighing of evidence based on personal probabilities. The human mind is often better than a computer at quantifying unclear evidence to reach a rational decision. The computer's 'best' is constrained by the ability of those who formulate the mathematical model and the skill of its programmers. A doctor's instinct may be more appropriate than trying to formulate a mathematical or statistical model of rational behaviour and programming a computer to make decisions on that basis. Computer scientists are nevertheless striving to do this sort of thing in this and other spheres by designing so-called *expert systems*. There is indeed a whole branch of computer studies now referred to as *artificial intelligence.*

Probabilities enter, at least subconsciously, into nearly all medical diagnoses. A vivid example of their influence was the controversy in 1987 over sexual abuse of children in Cleveland, England. In that region diagnosed cases of alleged sexual abuse rose dramatically. This happened when a group of specialists introduced a new test and accepted a positive response as a near certain indication of abuse. Medical opinion was divided on the reliability of the test and other doctors assigned a lower probability to a positive response indicating abuse. These probabilities were of course personal or subjective. Compared with reported levels of abuse in other areas, the large number of alleged cases in Cleveland caused surprise: such a high regional departure from national norms would have a very low (objective) probability. Findings of a detailed enquiry were released in July 1988.

## Causes of disease

Danger from smoking tobacco is a long-running saga, the issue for long confused. When Christopher Columbus introduced, and later Sir Walter Raleigh reintroduced, tobacco to Europe it had endorsement from American Indians as a beneficial medicine. As recently as the 1930s, unlike the Government health warnings of today, cigarette advertisements extolled the soothing effect of tobacco upon the nerves. I recall a rather negative (was it defensive?) assertion from one maker that his offering 'does not affect the throat'.

Not all Raleigh's successors saw virtue in the leaf. The erratic Algernon Swinburne, while berating James I, assigned to him the one redeeming feature that 'he slit the throat of that blackguard Raleigh, who invented this filthy smoking'. Robert Burton was schizophrenic on the subject:

> Tobacco, divine, rare, superexcellent tobacco, which goes far beyond all their panaceas, potable gold, and philosopher's stones, a sovereign remedy to all diseases ... But, as it is commonly abused by most men, which take it as tinkers do to ale, 'tis a plague, a mischief, a violent purger of goods, lands, health, hellish, devilish, and damned tobacco, the ruin and overthrow of body and soul.

Enthusiastic Arthur Helps in 1859 wrote 'What a blessing this smoking is! Perhaps the greatest that we owe to the discovery of

America.' Rudyard Kipling may well have had in mind moles in the security services of his day when he wrote of the English:

For undemocratic reasons and for motives not of State,
They arrive at their conclusions – largely inarticulate.
Being void of self expression they confide their views to none;
But sometimes in a smoking room, one learns why things were done.

To Oscar Wilde, a cigarette was 'The perfect type of a perfect pleasure. It is exquisite, and it leaves one unsatisfied. What more can one want?'

## A hint of danger

Victorian physicians occasionally suggested smoking might be a threat to health, but until this century condemnation was based on impressions, with no scientific evaluation of evidence. About fifty years ago, came suggestions from British and American doctors that nearly every lung cancer patient they saw was a smoker. In the mid-1930s two doctors claimed 90% of 135 men known to them to be afflicted with lung cancer were 'chronic smokers'.

Their report typified a weakness of some of the earlier evidence of a link between smoking and lung cancer. Their study was *retrospective*. They started with the event *patient has lung cancer*, then *looked back* to see how many of the subjects had the characteristic *heavy smoker*.

This approach is open to the criticism that, even if the great majority of those who contract lung cancer are heavy smokers, this is not evidence that smoking is the cause. Might not there be some other factor that predisposes people who would in any case develop lung cancer to smoke heavily? For example, if excessive tensions at work or whatever induce lung cancer (as indeed tension does certain forms of stomach ulcers), it would seem likely that potential victims of lung cancer might turn to smoking to relieve these tensions.

## Experimental restrictions

Could we do a Salk-type experiment on smoking? The second Salk test was one in which all factors other than that under investigation (administration of vaccine or a placebo) were kept as constant as possible. Also the likelihood of bias was reduced by random allocation of subjects to receive vaccine or placebo. It was a randomized clinical trial (a *double blind* because neither doctor nor patient knew who got vaccine or placebo).

Lung cancer, like polio, is a distressing disease that is not very common, so we would need a large experiment to test its association with a factor like smoking. However, a Salk-type experiment where a large number of people are selected and randomly allocated *either* to a group not allowed to smoke *or* to a second group instructed to smoke, say, fifty cigarettes a day was not on. It would be unethical – and illegal. We cannot expose people to a risk of lung cancer when we *believe* (even if it is not proven) that smoking may cause it. The situation with the Salk vaccine was different; there was at the time no evidence that it might be harmful (even though that later incident (p. 122) did indeed indicate a problem).

The best we can do is to carry out either retrospective – or, better, prospective studies – with people selected from those available in two or more categories: these categories might be non-smoker, moderate smoker, heavy smoker.

In 1939 a retrospective study was made selecting patients who were well (or suffered diseases other than lung cancer) and who were as like as possible in characteristics such as age, sex, rural or urban environment, etc., as some lung cancer patients. In these studies more smokers were found among lung cancer patients. The results were open to criticism for possible biases in selecting the control (non-lung-cancer) group. It is also difficult to classify people by smoking habits. Data may not be readily available, different people have different ideas about what is meant by 'heavy smoker' and 'moderate smoker'. Some once-heavy smokers may have given up; evidence about how long people have been smoking may be scant.

Studies like this were not enough to condemn unequivocally, let alone ban, tobacco. The industry is (even today) of considerable economic importance and imposition of any ban or restriction has political and legal implications.

## The wisdom of Pearl

At Johns Hopkins University, eminent biometrician Raymond Pearl (1879–1940) took a broader approach. He did not confine his study to lung cancer, but compiled information about smoking habits and lifespan of *all* patients in his files. He found that while 65% of non-smokers survived to age sixty, only 45% of heavy smokers did. Indeed, for all ages over thirty, he found a higher proportion of moderate or non-smokers survived than did heavy smokers, non-smokers doing marginally better than moderate smokers. His study was *prospective*: both smokers and non-smokers were first identified, then their history examined to determine cause of death. His findings were still open to criticism for possible bias. Why were patients on his files? How good was the information about smoking? It could still be that people with a predisposition to lung cancer, or other life-shortening influences, tend also to smoke and people who have no such predisposition do not, or are less likely to, smoke.

## British doctors and smoking

In 1964 Richard Doll and A. Bradford Hill published the now classic paper "Mortality in relation to smoking: ten years' observation of British doctors". This was a prospective study.

They sent a simple questionnaire to 60 000 members of the U.K. medical profession (of both sexes). About 40 000 responded. They studied the survival of these over ten years. They found the death rate from lung cancer among heavy smokers to be 1.66 per thousand compared with 0.07 per thousand for non-smokers. Thus, heavy smokers had a death rate from lung cancer some twenty-four times higher than non-smokers.

Although the lung cancer finding received most publicity, a more important finding of the Doll and Hill study was a higher death rate from cardiac failure among smokers. Findings strikingly similar were made by two American statisticians about the same time. Indeed, the greatest number of surplus deaths among smokers occurred in the cardiac circulatory area (see p. 6) because this is a more common cause of death, this despite the death rate being only some 50% (1.5 times) higher from this cause in smokers as opposed to non-smokers.

Other studies in the 1960s confirmed these findings. Two eminent statisticians remained unsatisfied because the evidence was not based on randomized clinical trials with precautions to 'balance' other factors that might affect the outcome. Such trials have a key role in medical research. Many countries do not allow new drugs to be used until they have been thoroughly tested in this way. We indicated on p. 208 why such trials are impossible in the smoking/lung cancer situation.

### The objectors

Ronald Aylmer Fisher (1890–1962), statistician and geneticist, whose ideas (formulated originally in relation to agricultural experiments) lie at the heart of randomized clinical trials, was one objector. He argued, much to the comfort of the tobacco industry, that the evidence was only circumstantial and that, if persons with an hereditary tendency to smoke also had an hereditary predisposition to disease, one could get results of the sort demonstrated by both retrospective and prospective studies.

While a reasonable hypothesis, Fisher's theory was discredited when later evidence showed that there was a close relationship between smoking and disease, heavier smokers being more at risk; also, those who gave up smoking had a better chance of survival. We shall see in our next section on genetical evidence that factors other than genetics clearly affect susceptibility to respiratory disease.

Joseph Berkson, the other eminent objector, considered it unlikely that the death rate from a number of different, and quite unrelated, diseases should all be increased by smoking, especially as there was no clinical evidence of how smoking damaged the body system.

Such evidence is now accumulating. Smoking has some effect on blood vessels in both animals and humans. We also know that in many people it reduces the respiratory rate and that this in turn can lead to infections of a bronchitis type (bronchitis is a medically ill-defined term covering a number of chest ailments). Peto (1975) reported that, if people whose respiratory rate was reduced by smoking to a level resulting in disablement then gave up smoking, this did not improve the respiratory rate but did halt the decline and increased life expectation by five or more years.

Further, a tobacco smoke environment has been shown to produce lesions and lung cancer in animals. This evidence, together with detection of carcinogens in tobacco, provide strong evidence that smoking is an important cause of lung cancer.

Few scientists subscribe to a pure 'causal theory'. Most admit that genetic factors (see also p. 213) could also influence incidence of lung cancer and some other smoking-induced illnesses.

The case against a causal link is still argued. P. R. J. Burch has written extensively on problems of inferring cause in this area and gives a somewhat technical presentation in Burch (1978). The ensuing discussion published with the paper is informative and easier reading for the non-specialist than the paper itself.

## Nature v. nurture

A study of twins may help decide whether a disease is caused almost exclusively by genetic factors or by environmental ones, or perhaps a mixture.

We can do this because twins are of two kinds – *identical* and *non-identical*. Identical twins come from the same fertilized egg; thus, each twin has the same gene set to determine physical and mental characteristics. This means each member of an identical twin pair has the same susceptibility to any genetically transmitted disease. Non-identical twins result from fertilization of separate eggs and are no more alike genetically than ordinary brothers and sisters in so far as they may or may not have inherited a gene with a particular characteristic.

If one of a pair of twins develops a genetically transmitted disease, it is more likely to appear in his 'pair' if the twins are identical (because each has the same genetic composition) than would be the case if the twins are non-identical (because the gene composition of each will have *some* differences).

*Table 12.2.* Incidence of a disease in twins ( × = affected, ○ = unaffected)

| Pair number | First twin | Second twin |
|:---:|:---:|:---:|
| 1 | × | × |
| 2 | × | × |
| 3 | × | × |
| 4 | × | × |
| 5 | × | × |
| 6 | × | × |
| 7 | ○ | × |
| 8 | × | ○ |
| 9 | × | ○ |
| 10 | × | ○ |

Reid (1972) discusses an historic study of Danish twins born between 1870 and 1910. It was carried out on 4368 pairs, where each twin was of the same sex and it was known whether or not each pair was identical. A measure frequently used for the concordance between cases of a specific disease in twins is the ratio of the number of cases in twin sets where each is affected (i.e. both twins have the disease) to the total number of cases among all twins with at least one case. This is usually multiplied by one hundred and expressed as a percentage called the *affected concordance rate*.

For example, if we have, as in Table 12.2, ten pairs of twins with at least one member affected by a disease and in six of these both twins have the disease, then there are sixteen cases altogether and twelve of these are 'paired' (i.e. it occurs in each of six twin pairs); for this data the *affected concordance rate* is therefore $\frac{12}{16} \times 100 = 75$.

Since a genetically transmitted disease is more likely to appear pairwise in identical twins, this implies that, if we calculate an *affected concordance rate* both for sets of identical twins and for sets of non-identical twins, then that for the former will usually be appreciably higher. If the disease is one of environment, we would expect the rates to be much the same for identical and for non-identical twins.

For the Danish twins, the *affected concordance rates* for cerebral apoplexy, for example, were thirty-six for identical twins and nineteen for non-identical twins. A difference this size suggests a genetic link. On the other hand, for deaths from acute infections, the rates were, respectively, fourteen and eleven. The closeness of these rates suggests little or no genetic link.

Because lung cancer deaths are rare, the Danish data are not sufficient to test for a genetic link. However, they have been used to see if there is a genetic link between smoking or not smoking. It turns out there is, the evidence showing that, if one of a set of twins smokes, his fraternal twin is more likely to do so if they are identical twins than if they are non-identical.

While we cannot see if this has relevance to lung cancer (because of small numbers), we can look at the situation for bronchitis. What one does is look at twin pairs in which one twin smokes and the other does not. If the twins are identical and bronchitis is caused by a genetical factor *unconnected with smoking*, one would expect about the same proportion to suffer from bronchitis among members of the pairs who did not smoke as among members of these pairs who did smoke. The results show that for both sexes bronchitis is about twice as common for smokers as for non-smokers (the percentages were approximately 14% for smokers and 7% for non-smokers). The results were strikingly similar for non-identical twin pairs one of whom smoked while the other did not. To summarize, these results indicate that, while predisposition to smoke may have a genetic element, once one has decided to smoke, the danger of bronchitis is increased irrespective of this genetic factor.

It would be interesting if we had results for a larger twin study to see if the same held for lung cancer and heart disease. The result for bronchitis can at best be described as a significant pointer that smoking can affect health independently of genetic factors.

## Extrapolation dangers

Care is needed in extrapolating from twin studies to the population at large, for the following reasons:

(i) Only about one birth in ninety-six is twin.

(ii) Twins are more common in certain ethnic groups and with older mothers; this is relevant because some diseases are more common in certain races, others in the offspring of an older mother.

(iii) Twins are often brought up in a similar environment, at least as children, so environmental factors may be mistaken for genetic ones. This is less likely for adult twins.

(iv) Twins have to compete for resources in the womb so may be less favoured, at least in early life, than single-birth offspring; this is most likely to affect infant morbidity and mortality and to be less important in later years.

## Screening

Successful treatment of many diseases – glaucoma, cervical cancer and breast cancer – is helped by early diagnosis. Screening techniques exist to detect these conditions before clinical symptoms appear.

Optimum screening policy depends on several factors. As well as effectiveness, safety and cost, we have to remember the human angle, including the reaction of individuals to findings, positive or negative, from a screening. There is an interplay here between economic and statistical desiderata and medical, ethical, humanitarian and social issues.

Participants and organizers may perceive differently the purpose of a screening programme. People may be reluctant, even refuse, to participate either because (i) they feel they are not at risk or (ii) they regard the scheme as an intrusion on their rights. The organizers will almost certainly regard the project as a worthy attempt to reduce or eradicate a serious or distressing illness.

Good screening programmes are evolutionary, both to keep up with technology and because operating the scheme may indicate hidden dangers due, for example, to radiation.

Use of X-rays to detect breast cancer – mammography – carries a small danger from radiation, but new technology minimizes this if screening is not too frequent. Mammography is important because it detects at an early stage (when treatment is less drastic and more effective) nearly half as many cases again as do other methods.

## How often?

For any disease, how often to screen depends on the following factors:

(i) the 'lead time' between the earliest instant at which the screening picks up the disease and the appearance of clinical symptoms;
(ii) how important it is to make an early diagnosis.

For breast cancer, early detection and modern treatments avoid disfiguring surgery. Determining optimal policy is difficult: lead time varies between individuals and may also be a function of age. False negatives are another problem, so an associated clinical examination is desirable as some of the cases missed by mammography may be detectable on examination.

Mathematical models have been proposed to determine an ideal interval between screening. These depend upon assumptions regarding how the disease progresses, so there is a certain pragmatism in deciding frequency of screening. Intervals of two to three years are commonly recommended.

## The New York study

Goldberg and Wittes (1981) describe a New York study in the 1970s. A total of 132 cases were detected: fifty-nine by clinical examination only; forty-four by mammography only; and the remaining twenty-nine by both methods.

A total of 31 000 women were invited to take part. Some refused; others attended only one or two of three possible screenings. Of those who refused, seventy-four later developed the disease. For them, diagnosis was made by medical examination, often after the patient herself had noticed a lump,

False positives had a deterring influence, causing unnecessary fear. People incorrectly screened positive did not always attend later screenings, or ignored a later true positive.

## Who attends?

With voluntary screening, those attending may not be typical of the population at risk (e.g. for cancer, there may be a preponderance of volunteers from families with a history of breast cancer) or from particular ethnic or social groups. In the New York study an invitation was more likely to be taken up by younger, married, Jewish and better-educated citizens.

Simple considerations like location of screening centres and pleasantness of surroundings and friendliness of staff influence the likelihood that people return for routine follow ups. As screening is often for low probability events, if it is unpleasant, people may neglect a small risk; we have a good example here of a personal, but not necessarily rational, assessment of risk–benefit balance, of utilities.

How does heightened awareness influence people's attitude to attending screenings? Does appearance of a suspicious symptom encourage a person to attend in the hope of getting early treatment? Or does fear of the finding discourage them? Publicity (in the 1970s) when a U.S. President's wife developed breast cancer (successfully treated) temporarily increased demand for screening, with a decline once the publicity faded. There was also a decline in the U.S.A. after suggestions that mammography carried radiation dangers. Will publicity in 1987 for yet another President's wife being treated for breast cancer again increase demand for screening?

## How effective?

Several ways have been proposed to assess effectiveness of screening. Blumenson (1976) suggested measuring the fractional decrease in numbers of undiscovered (by the screening) non-curable cases when a screened population is compared with a non-screened one. On this basis, he found little advantage in screening women for breast cancer at age forty rather than at age fifty. Other measures of effectiveness are based on comparison of mortality rates between screened and unscreened populations. An important question here is whether cases detected by screening progress at the same rate as those detected by usual methods; they may not if, for example, a higher proportion from families with a history of breast cancer present themselves. Also, if some cases regress (as cancers sometimes do), screening may detect some (regressive) cases that would be missed by other methods, as the patient in these circumstances would, in fact, recover naturally without need for treatment.

### Drugs and danger

In the U.K. and other countries there are licensing or similar requirements before drugs can be put on the open market or made generally available on prescription. Usually they are tested first on volunteers (even earlier on animals), then on 1000 or more selected and carefully monitored patients in clinical trials. All side effects are recorded. If there is evidence of a dangerous adverse reaction, the drug will almost certainly be withdrawn at this pre-marketing stage. In the U.K. licences are issued on the advice of the U.K. Committee on the Safety of Medicines after it is satisfied that the drug has therapeutic benefits and any recognized side effects are unimportant relative to the potential benefits.

A difficulty is that serious side effects may occur only for a few patients, perhaps only one or two in about 10 000. Also, they may only appear after the drug has been in use for a long period, or even years after a patient has stopped using the drug.

### *The yellow card system*

To alert the Committee on the Safety of Medicines to possible effects, there has operated in the U.K. for some twenty years what is known as the *adverse reaction reporting system*. All doctors are asked to fill in a form (called the yellow card) and submit it to the Committee if, after prescribing any drug, they notice an adverse reaction which might conceivably be related to the drug. There is a subjective element in what constitutes such an effect, but most doctors are conscientious about reporting, and in 1986 over 15 500 reports were received by the Committee. The scheme covers *all* drugs and *all* patients. The system may give early warning of possible dangers. If a drug is suspect on other grounds (e.g. experience in other countries of known problems with related drugs), yellow card reports may confirm these suspicions.

## The green form

A second scheme in the U.K. is the *prescription event monitoring system*. Complementary to the yellow card scheme, it operates for a limited range of drugs which for various reasons (novelty, mode of action, etc.) are thought *might* carry a risk. Under the scheme, doctors prescribing any of these drugs notify the body running the scheme. One year later, the doctor is asked to fill in a green form reporting any 'event' in the patient's medical history, *whether or not* it seems related to use of the drug. Unlike the yellow cards, which only report suspicion of a side effect, the monitoring system covers most patients receiving the relevant drug and reports all items of medical history that may or may not be related to drug action. In this way, unexpected side effects may become apparent.

## Detective work

Perversely, it is often easier to pick up a side effect that produces a few extra cases of a very rare disease than one producing a larger number of cases of a more common disorder. Here is an example based on a real case (with the data slightly simplified for illustrative purposes) for a drug that was eventually withdrawn following incident reports after it had been prescribed in 10 000 cases. The drug was not for treating bone-marrow disorders and it was known that the normal incidence rate for the bone-marrow disorder *aplastic anaemia* was one per 20 000. Among the 10 000 users of the drug, five cases of *aplastic anaemia* were noted. Now, the 95% range (p. 93) for a normal incidence rate of one per

20 000 is, for a sample of 10 000, from zero to two. Since five is outside this range, the result is significant. It is, indeed, also outside the relevant 99% range, implying a highly significant increase. If this rate were maintained, it would suggest an additional 4.5 deaths per 10 000 people taking the drug.

For acute *myocardial infarction*, the normal incidence rate is fifty per 10 000. For a sample of 10 000, appropriate calculations give a 95% range from thirty-seven to sixty-four. Among 10 000 taking the drug (which was not for treating *myocardial infarction*) there were sixty cases, so, using the 95% range criteria in a hypothesis test, we would not reject the null hypothesis that the incidence rate is still fifty. Yet, if the increase to sixty were maintained, this would mean the drug is giving ten additional cases per 10 000 users, compared with 4.5 additional cases of *aplastic anaemia*. Indeed, to establish firmly that there is a real increase in *myocardial infarction*, we would have to continue using the drug until more evidence accumulated. This clearly demonstrates an unpleasant difficulty with some uncertainty situations. There may well be a difference that is not *statistically significant*, but which is none the less of *practical importance*. This dilemma has to be faced with any quantitative assessment of risk.

## The media kill

Most drugs have some side effects; fortunately many are only minor irritations. Certain individuals may show allergic reactions to groups of drugs. Doctors are well aware of this, and will avoid or stop prescribing a drug in these circumstances. Just what side effects are tolerable depends on the benefits bestowed by a drug and the availability of alternatives, as well as on the severity and frequency of side effects. Many drugs used to treat some types of cancer have extremely unpleasant side effects. If these are not too long lasting or permanently debilitating, patients are usually willing to put up with them if the alternative is death. The Committee on the Safety of Medicines weigh carefully such matters in advising whether or not a drug should be marketed or withdrawn.

Unfortunately, the media, hungry for a good news story, are

sometimes less scrupulous. They may report *one* authentic case of a serious side effect of a widely used drug, embellishing it with unsubstantiated suggestions that many doctors are calling for restrictions on its use. Such evidence is often hearsay picked up by one journalist talking to one doctor; there are eccentrics in most professions.

Media reports on these lines are very worrying for those taking the drug. There is at least one confirmed case in the U.K. where a Sunday paper published just such a scare story. A patient for whom the drug had been prescribed as a 'life-saver', alarmed by the report, immediately stopped taking the drug. She was not alive to read her paper the following Sunday. This is a clear case of a media killing: what the poor patient did not realize – because it was not clear from the media report – was that the side effect would probably only occur in less than one case in 1000, that it was serious but treatable and that her life depended on continuing to take the drug.

Responsible reporting of drug dangers is to be encouraged. Sadly few journalists – even many so-called 'medical' or 'science' correspondents of our newspapers – have the ability or information needed to make an objective assessment of such dangers. Even if they do, they are unlikely to be given the press space or broadcasting time needed for a full informative report. The Committee on the Safety of Medicines, a body of the top experts in the U.K. on drug therapy, find their task a daunting one, not helped by media-style scaremongering.

## A review

The medical field brings together all the aspects of uncertainty met in the earlier chapters of Part II. In diagnosis subjective probabilities and utilities are important. In assessing these, a doctor will also take objective probabilities into account – the relative prevalence of different diseases and how often certain symptoms occur with various diseases.

Classical statistical studies founded on probability theory may be used in studies of causes of disease (smoking) or in testing the safety and efficacy of drugs. We may use banks of data (the Danish twins) to make deductions about genetic links with disease.

Screening programmes involve assessments of risks such as that from radiation and an appreciation of the psychological reactions to discomfort in the process, or to potential findings, especially false positives.

The ever-present danger of serious side effects, especially if rare, poses problems with drug usage and control. Here, the U.K. yellow card and green form procedures are examples of sensible ways to collect relevant data to alert us to risks, emphasizing the need for sound, and often far-reaching, data collection methods.

# CHAPTER 13

# STATISTICS AND THE LAW

## Reasonable doubt

Criminal law requires that guilt be established *beyond reasonable doubt*. This clearly has probabilistic overtones, so it is ironic that only recently have courts regularly considered probabilistic evidence in the scientific sense.

Despite its probabilistic connotation, *beyond reasonable doubt* is not assigned a numerical value; it might be unwise, impossible even, to do so. Some feel the degree of doubt should not be divorced from the seriousness of the crime or the consequent sentence if a person be found guilty. But how? Do we argue that the more serious the charge the higher the degree of *proof* we should require? Or agree with those who believe that, if, say, an accused murderer is acquitted because of a mere shadow of doubt, this is unfair to society, as he might well commit a similar offence, whereas it matters less if a first-offender correctly charged with being drunk and disorderly avoids conviction?

Civil disputes are usually adjudicated on a *balance of probabilities*. A lawyer may not think of this in numerical terms, but the critical concept here is whether a probability is greater or less than one-half (or the *odds against* greater or less than one).

Despite these probabilistic elements in their decision making, the legal profession and courts are often wary about calling for or accepting the evidence of expert statistical witnesses, a reluctance in part due to problems of communication between statisticians and lawyers. Both must share the blame.

## Rules of evidence

Court procedures, in particular the British adversarial, as distinct from inquisitorial, approach to trials, can militate against a fair presentation of quantitative evidence. Legal decisions are arrived at by inductive arguments, attempting to establish or disprove an accused's guilt by examining the implications of evidence. A consequence is that it is hard to avoid distortions or the making of false inferences by a skilled attorney. The approach contrasts with that of logical deductions that are the essence of the search for scientific truth.

Occasionally a statistical difficulty may face a witness with statistical knowledge who is not appearing as an expert. This can be frustrating. I was called as a witness for the prosecution in a case arising from a traffic accident. The defence asked me to agree that a plan represented the scene of the accident *approximately*. 'Do you agree that the key features shown are approximately – to within a few feet, say – in their correct relative positions?' I was asked. My answer was yes. What was not brought out by question or answer was that the distortions all favoured the case of the defendant: walls moved to obscure his vision, a post shifted slightly to imply another vehicle would be hidden from the defendant's driving position for a longer time than it was, and so on. As a non-expert witness, the rules of evidence did not allow me to express this 'opinion'. The prosecuting lawyer did not appreciate the nature of the distortions. Here was a clear case of uncertainty being used to the advantage of one party.

Had I been called as an expert on statistics, I would have been free to give an *expert opinion* on the nature and effect of these distortions. Ordinary witnesses must adhere to matters of fact, not express opinions or present hearsay, restrictions widely felt to be in the interests of justice. Usually they are: there must be few cases where a witness called to describe what he knows of the circumstances of a case turns out, by accident, to be expert on an *incidental* feature of the evidence. However, proposed changes in English law may remove some restrictions on admissible evidence in the near future.

## The truth, the half truth . . .

A layman tends to equate justice to truth. But, as Napley (1982)
points out, the Court, for reasons of economy of time, is only
interested in aspects of the truth needed to establish justice. For
example, if A is accused of murdering B, the court is only interested
in whether or not A committed the crime, not in how or why,
unless this is essential to establishing that he *did* commit it. With
widespread criticism of delays in the courts, this restriction is
possibly justified, though the layman may regard knowing the
whole truth as relevant to ensuring that *the punishment fits the
crime*.

### Statistical expertise in a legal context

The late Frank Downton, an expert on statistical aspects of gaming
legislation, listed six fields where statisticians have a part as expert
witnesses or to assist in framing or interpreting laws (Downton,
1982). These were as follows.

1. *Non-numerical uncertainty*. Can a survey provide valid conclusions? A manufacturer of microwave ovens in the U.S.A. claimed
a survey showed *independent* service technicians preferred that
maker's ovens to other brands for a number of reasons. Statisticians advised that the survey did not substantiate the claim
because it included only dealers for the particular make and other
authorized service agents, excluding those who serviced the ovens
but were not factory authorized to do so (and therefore really
*independent*). They made other less important criticisms. Advice
may also be given on bias in questions or effects of non-response.

2. *Framing and advising on legislation*. On p. 233 we explain the
statistician's role in framing and administering weights and measures legislation; that law could not work without a statistical
input. Gaming laws also have an essential statistical component,
so may Clean Air Acts and similar anti-pollution legislation. Statisticians have even been asked to examine the workings of the law
itself. Do judges and juries commonly agree? Do six-man juries
(used in some U.S. cases) come to similar decisions to twelve-man

juries and, if so, are there advantages (timewise and costwise) in smaller juries? The statistician also has a role in interpreting criminal statistics in relation to assessing police policy and administration of justice.

3. *Explaining actuarial data.* Accident compensation claims are often settled on the basis of expected loss of working life, etc., based on actuarial tables. The legal profession usually understands the notion of *expected* life, but show little appreciation of individual or group variation. An actuary or statistician may help by explaining how to interpret life tables, by, for example, drawing attention to the different life expectations of a coalminer, a middle-rank civil servant or a university don, and to variation about the expected lifespan for any individual.

4. *Interpretation of small sample statistics.* If, in a town, hotel closing hours are changed from 10 p.m. to 11 p.m., it might be argued that the change was beneficial because there were less street arrests in the hour following hotel closing time after the change than was the case before. An appealing argument, but a statistician would urge caution in accepting it, because (i) it may not be clear how many street arrests involve people who have emerged from hotels at or near closing time; (ii) numbers on the streets before 10 p.m. and 11 p.m. (relevant hour before the change) may not be the same as between 11 p.m. and midnight (relevant hour after), so the potential numbers *available for arrest* may be different; and (iii) we are considering figures for different years, and police policy towards street arrests may have changed. These points should be examined to assess the relevance (if any) of numbers of arrests to arguments for or against later hotel closing.

5. *Applied probability.* This is often relevant in forensic evidence. Glass fragments similar to those at the scene of a crime are found in a suspect's clothing; the evidence is more convincing if the type of glass is rare than if it were a common window glass or the kind used for milk bottles. It would be even more convincing if both samples of glass were smeared with an unusual type of grease. Downton quotes an interesting example using probabilities. A man was found in possession of 526 Seiko watches. At the time,

there were no less than 715 types of watch of this make in circulation. A missing batch from an airport warehouse contained 741 watches of fifty-three different types. The numbers of each of the fifty-three types were known. The 526 watches found in the man's possession consisted of forty-eight of the fifty-three types in the missing batch (and no others), all in numbers equal to or less than those in the missing batch. The court was told that the probability of such a coincidence was less than one in $6 \times 10^{28}$. This calculation depended on certain assumptions which might be disputed. For example, if Seiko regularly packed batches of the size of the missing one, containing the same numbers of each kind of watch, then the watches in the man's possession might well have come from *any* such batch. Of course, had the number of any kind in his possession been greater than the number in the missing batch, or had there been watches of a kind not in the missing batch, then clearly not *all* watches in the man's possession could have come from that missing batch.

We discuss on p. 228 the case of *People v. Collins*, a classic demonstration of probability pitfalls.

6. *Pure probability*. This is especially relevant to gaming legislation and to settling arguments about breaches of such law, involving, for example, false claims about odds. Gambling has come a long way since de Méré's day (p. 55)!

### Expert witnesses in action

One problem familiar to expert statistical witnesses is the difficulty that some lawyers, judges and 'juries have in understanding numerical evidence, often resulting in misinterpretation.

Newell (1982) gives a lively account of the difficulties statisticians have in communicating with lawyers. Not least is the latter's (understandable) inability to appreciate technical terms and a tendency by lawyers to confuse, accidentally, or even deliberately, the witness on these matters in a way that produces irrelevant answers.

Newell was involved in a long case about fluoridation of water. It lasted 143 days and he was only discharged as a witness on the

119th day. The case produced convoluted questions. An examplе:

> Q. Well if urbanization is irrelevant in the static situation, and if the difference in racial structure as affecting cancer death rates is caused by urbanization, then it follows that the difference of racial structure as affecting cancer death rates is also irrelevant?

The court spent one and three-quarter hours sorting that one out and getting an answer: nearly sixty pages of transcript to record all this. Newell's paper spells out how the discourse developed.

Incidentally, draft court transcripts do not always record technical matters accurately and unless errors are quickly corrected they may, especially in a long case, be taken later as true versions of evidence. There are dangers though in allowing corrections to transcripts. Wittingly, or otherwise, people may try to get transcripts amended to *what they meant to say* when the record is in fact *what they did say*. Newell suggests that, for technical matters, tape recordings may be better.

### Conflicting evidence

Since adversaries may each call their own experts, judges may find themselves in the difficult position of having to arbitrate between expert opinions in a field where they lack, and cannot be expected to have, expertise. Newell suggests statistical (or other expert) assessors might sit with a judge, when necessary, to arbitrate between experts. The law varies from country to country (even between England and Scotland!) on permissibility of such assessors.

How are expert witnesses selected? They are seldom called as witnesses without having previously advised, however informally, the lawyer for the side on which they appear. If that lawyer does not think the expert's opinion will help his client's case, the expert is unlikely to be called (except perhaps by the other side!).

Even if he makes it to the witness stand, the impartial scientific approach of a statistician may appear to the court to be prevarication. Statisticians often assert a relationship exists, yet are reluctant to attribute cause and effect. In Chapters 7 (the Salk vaccine) and 12 (smoking and health) we saw some of the problems in confirming causal chains; this, not prevarication, is the usual reason for such reluctance.

## The communications gap

As perplexing as the convoluted question is the unanswerable one, so phrased because a lawyer does not understand a technical point. An extreme example quoted by Newell was:

> Q. The chance of getting four heads is 0.06. So the chance of getting $3\frac{1}{2}$ heads?

Lawyers, too, not uncommonly misuse technical terms in a way that makes it clear they do not understand statistical reasoning. The statistician is doubtless to blame in some cases by using everyday terms such as *significance* with a specific technical meaning. Surprisingly, lawyers are often keener to learn how a technique is carried out than in establishing what a test or estimation procedure is intended to achieve or whether it is relevant to a particular matter.

## A classic case of (mis)use of probabilities

The American trial *People v. Collins* has been widely reported and discussed; probabilities played a key role. A Californian jury found Malcolm Ricardo Collins, a coloured man, and his Caucasian wife, Janet Louise, guilty of second-degree robbery. The alleged incident took place on 18 June 1964.

The evidence was conflicting: the prosecution relied heavily on the testimony of a college mathematics lecturer regarding the probability that a couple in the population at large had the characteristics of the accused pair. The jury were asked to believe that, if this probability was small, and a couple with these characteristics was seen at the scene of the crime, then it was highly probable (i.e. beyond all reasonable doubt) it was the accused who committed the crime.

*Table 13.1.* Probabilities of characteristics in
*People v. Collins*

| Characteristic | Probability |
| --- | --- |
| Partly yellow automobile | $\frac{1}{10}$ |
| Man with moustache | $\frac{1}{4}$ |
| Girl with pony tail | $\frac{1}{10}$ |
| Girl with blonde hair | $\frac{1}{3}$ |
| Negro man with beard | $\frac{1}{10}$ |
| Interracial couple in car | $\frac{1}{1000}$ |

Collins, but not his wife, appealed. The court quashed the conviction and ordered a retrial on the grounds that the statistical evidence was not satisfactory. The appeal decision was clearly influenced by three beliefs:

(i) The probabilities of individual outcomes (events) used in the evidence (discussed below) were not arrived at logically.
(ii) They were not validly used to calculate the overall probability associated with random selection of a couple.
(iii) Even had the probabilities been valid, the inferences drawn from them were not.

At the original trial the probabilities in Table 13.1 were suggested by the prosecuting attorney as 'reasonable'. The college lecturer called as a statistical expert agreed.

These characteristics were said by witnesses to fit a couple seen near the scene of the crime, though not all witnesses described all characteristics. No evidence was given as to how the probabilities in Table 13.1 were arrived at, prosecuting counsel even inviting jury members to substitute their own assessments of probabilities for these events. The expert gave evidence that 'the probability that any couple possessed the distinctive characteristics of the defendants' was $\frac{1}{12\,000\,000}$, the product of all probabilities in Table 13.1. From this, the jury were invited to infer that it was 'virtually impossible' for another couple with these characteristics to be involved. The jury appeared to accept this, though other evidence may have been relevant to their conclusion.

## Determining probabilities

How was that probability of 1 in 1000 that an interracial couple
share a car arrived at? An inspired guess, or from a sample survey?
The court were not told. Nor were they told if observation confirmed one car in ten was 'partly yellow'. To do so would first
require some refinement in definition. Does *partly yellow* in fact
mean *predominantly yellow* – the evidence pointed that way. If so,
is the proportion reasonable? The point (ii) raised in the appeal
judgement (p. 229) is also important. If these probabilities are
accepted, is use of the multiplication rule justified? Remember,
there are two forms for the multiplication rule, depending upon
whether or not events are independent (pp. 61–62).

The appeal court concluded that *on the record before us defendant
should not have had his guilt determined by the odds and that he is
entitled to a new trial.*

The judges commented on each point (i) to (iii). Quite rightly,
they questioned the probabilities in Table 13.1. How were they
obtained? To what population did they apply? Maybe, in 1964,
one in four men in California had a moustache. This could have
been checked by a survey, calculating a relative frequency.

In multiplying probabilities did the lecturer assume the events
were independent, or that the probabilities given were conditional? In an informative dialogue between a supposed lawyer
and statistician, Fairley and Mosteller (1974) pointed out that the
appeal court did not seem to be aware that the probabilities might
be conditional, nor that multiplication would be permissible if
they were a set of conditional probabilities ordered in some way
to indicate what was conditional upon what. Fairley and Mosteller
suggested working from the bottom up assuming appropriate conditioning. Perhaps the appeal judges took it as self-evident that
the given probabilities could not be conditional. These are technica-
lities they could not be expected to appreciate, and a statistical
witness should make it clear whether he is dealing with conditional probabilities and, if so, on what events he is conditioning.

## Inferences

It was wrong to imply, as the prosecution did in the original hearing, that, because the probability a randomly selected couple from some unspecified population had the given characteristics was low, then the couple arrested *must* have been the ones present at the scene of the crime (and by further implication guilty). What is the relevant set of potential criminals? If we accept the evidence of eye-witnesses on characteristics, suspects are confined to people having these characteristics. There may have been a few other couples in the Los Angeles area who also fitted the bill; if so, all are equally suspect on this evidence alone. Some may be ruled out by satisfactory alibis, or for other good reasons; meantime they are at least as suspect as the Collinses.

A practical problem is that the police may not have records of all couples with these characteristics. In their discussion Fairley and Mosteller point out that, if there is one couple in a fairly large population with the characteristics of the accused, then the probability of their being another may be as high as 0.4. Not negligibly small, as the judge and jury appeared to believe in the original Collins trial.

Criminality is associated with socio-economic factors. Is it not possible that in Los Angeles negroes have a predilection to yellow cars? That, if they have criminal tendencies, they often grow beards and have liaisons with pretty Caucasian blondes? I have no idea if this is so, but, if there were a dozen couples in this category in Los Angeles, it surely would have been appropriate for the police to consider them all before making an arrest.

There was one final thrust from the appeal judges, who stated:

> While we discern no inherent incompatibility between the disciplines of law and mathematics and intend no general disapproval or disparagement of the latter as an auxiliary in the fact finding process of the former, we cannot uphold the technique employed in the instant case.

One must agree. They added with the typical fear that the non-numerate have for mathematics:

Mathematics, a veritable sorcerer in our computerized society, while assisting the trier of fact in the search for truth, must not cast a spell over him.

This it seemed to do in the original Collins hearing.

## Race and crime

We mentioned on p. 225 the statistician's role in seeing that crime statistics are carefully interpreted. Peculiarities of race and crime statistics have recently been highlighted by Monica A. Walker, an expert on criminological data (Walker, 1987).

In London in 1983, 17% of all people arrested were black (of Afro-Caribbean origin, but excluding Asiatics from the Indian subcontinent), while only 5% of the population were black. For the particular category *robbery and other violent thefts*, 42% of those arrested were black.

Two obvious but completely opposed interpretations are:

 (i) the crime rate is higher among black people;
(ii) the police 'have it in' for blacks and are more likely to arrest them.

Monica Walker takes a more balanced view, concluding that, with available information,

> Valid conclusions about different offender rates cannot be drawn unless the data are analysed in relation to the age, social class and other factors that have been found to relate to criminality. Besides this, the very low reporting, recording and clear-up rates (which may vary between these groups) may distort the final pattern of offender rates.

The details are in Walker's clearly written paper, but some points she makes are:

(i) The two groups may have the same offence rate, but an offence by a black person may be more likely to lead to an arrest than a similar offence by a white person. It is estimated that in London only about two people are arrested for every hundred robberies. However, if the offence rates for robbery were the same for blacks and whites, the figures imply the arrest rate for blacks must be fourteen times higher than that for whites. This could (but there is no proof that it does) arise because people are more inclined to report a robbery by a black person, or because black people are less good at avoiding arrest. Many arrests follow the police stopping people in the street and questioning them. Do black people spend more time on the streets?

(ii) If crime rates for the two groups are indeed different, this might be a reflection of relative numbers in different social classes or age groups: it is known that there is a higher proportion of offenders than in the population as a whole among manual workers, the unemployed, the socially deprived and the young. There are a higher proportion of blacks than whites in these categories.

Monica Walker also stresses that skin colour is an unsatisfactory criterion for ethnic grouping and that colour may be difficult to record (especially for crimes at night). It is possible, too, that police attitudes towards arrests are different in areas where particular ethnic groups predominate.

One can only endorse Walker's plea for more information before concluding either (i) that some ethnic groups contain a larger proportion of criminals than others or (ii) that the police 'have it in' for some groups.

### Weights and measures

Our final example shows the role of statistics in framing legislation. A detailed description of statistical aspects of the U.K. Weights and Measures Act, 1979, and the ensuing regulations, is given by Bissell and Pridmore (1981).

The legislation meets E.E.C. requirements and, unlike earlier U.K. legislation expressing weights of packages in terms of a guaranteed minimum, the act requires contents to have a given 'average' weight. Tolerance is allowed for slight deficiencies in a small proportion of packs.

Statisticians had two roles in 'marketing' the new legislation. First, to see that statistical concepts were properly defined in the Act. One must obey the letter of the law, so, if a piece of statistical nonsense creeps in, it must be obeyed even if it is nonsense (at least until the law is amended). In the original printing of regulations under the Act, an incorrect formula was given for calculating the sum of squares of deviations from the mean. Fortunately, this was noticed before the regulations became law.

Digressing momentarily, gaming laws are another area where the law may impose practical restrictions. Players of jackpot fruit machines will be aware that a player does not know whether he will be able to HOLD reels at the position they were in after the previous play until he has paid his stake money for the next game. This is a legal requirement in the U.K. because the 1968 Gaming Act is framed in terms of *the charge for playing a game once* and *the value of the prize which can be won by playing a game once*. The Gaming Board became aware of a difficulty in defining a single game if one could hold a winning line and so get a prize in another game. They ruled that, to comply with the Act, machines should be designed so that *the player has no way of knowing, until after he has committed himself to the game, whether or not he will be able to hold any of the reels at the positions at which they stand over from the previous game.*

### A guiding role

Returning to weights and measures, the statistician's second role was to help draft explanatory guides on how to test packages to see whether they complied with the regulations. These included a *Packers' Code* and *Inspectors' Manual*. Simple tests were needed that could be performed and interpreted unambiguously by non-experts.

Packers are required to carry out *reference tests* to see if goods measure up to specifications. Except for very small batches, the tests are performed on random samples from production runs (not unlike the samples taken for acceptance sampling (p. 178), but they are used in a different way). The sample size depends on the number in the production run.

## Deciding factors

For goods sold by weight, the crucial factors for deciding whether a batch is substandard are the declared average weight per package and the numbers in the sample that fall below a *tolerable negative error* (TNE). No item in a sample used in a reference test is allowed to have a weight that falls below the declared average by more than two TNEs; such packages are termed *inadequate*. It is an offence knowingly to market such a package. The TNEs for packages of different sizes depend on the average contents. For a package of average weight 100 g, it is 4.5 g; for a 1000 g package, it is 15 g. The test criteria depend, not only on weight per pack, but also on the size of the production batch. For a batch of 400, a sample of fifty must be taken. That sample will pass the reference tests if certain conditions hold. There are alternatives, but one acceptable set of conditions is as follows:

(i) If $Q_n$ is the declared average weight, not more than three of the sample of fifty must be 'non-standard', where non-standard means of weight less than $Q_n - \text{TNE}$, but not less than $Q_n - 2 \times \text{TNE}$.

(ii) The mean weight of the sample must be at least $Q_n - 0.503s$, where $s$ is the sample estimate of standard deviation. (How to calculate $s$ is defined in the regulations.)

The test criteria differ for batches and samples of different sizes. They are sensible if one assumes weights are normally distributed (p. 104). This is a reasonable practical assumption.

To ensure careful and accurate weighing of samples, there are limits to permitted errors in carrying out reference tests.

Special provisions are made if testing is necessarily destructive (e.g. involves opening of containers and removal of contents), and allowances are made for goods which lose weight through drying out during shelf-life.

Goods sold by volume rather than weight are tested in analogous ways. There are some exceptions to general rules: for example, goods sold in 'one-off lengths' (e.g. curtain or dress material sold by the metre) must be *at least* the nominal length.

## A review

There is a touch of irony about the reluctance of the law to come to terms with scientific aspects of uncertainty despite the probabilistic element implicit in the way 'just' decisions are arrived at.

The communication gap at expert-witness level is one reason for this reluctance. The gap is perhaps indicative of stubbornness in both professions. Other difficulties in giving due weight to statistical evidence are lawyer's irritation at a statistician's reluctance in assigning cause and effect and the problems faced by judges in assessing the merits of opposing views of expert witnesses in an area where the judge himself has no expertise (and cannot be expected to have it).

The formal role of the statistics in a legal framework (weights and measures, gambling legislation) is a comparatively recent development, and one likely to be of increasing importance as we come to terms politically with the inevitability of uncertainty.

It is perhaps in forensic matters that classical statistical methods are most widely used, and they are also likely to be employed increasingly in interpreting crime statistics.

# CHAPTER 14

# PATTERN RECOGNITION

## Some classic examples

Pattern recognition is basic to communication. Language attaches meaning to sounds and combinations of sounds, a multiplicity of tongues being an unfortunate consequence of different people attaching varied meanings to the same sound.

Written words are codes to convey meanings that correspond to spoken sounds. Ambiguity may creep into speech or writing, especially when words identical or similar in sound or spelling have more than one meaning. A simple adjustment to the code may remove ambiguity. To write 'I like four minute eggs for breakfast' is ambiguous. Does it mean an unspecified number of eggs boiled for four minutes, or four very small eggs? Insertion of a hyphen, 'I like four-minute eggs for breakfast', conveys the former meaning. If the other is intended, we do better to write 'I like four very small eggs for breakfast'. The hyphen is often important in writing: 'I own a little-used car' need not mean the same as 'I own a little used car'. The alternative meaning is implicit if we write 'I own a small used car'. Synonyms do more than give variety; they may remove ambiguity. Use of capital or lower case letters may imply subtle distinctions. A Christian scientist need not be a Christian Scientist. In speech we sometimes indicate different meanings by a pause or change in emphasis or pronunciation.

We can all recall examples where a written or spoken message is unclear. The 'double-entendre' is the stock-in trade of many comedians, the salvation of crossword puzzle compilers.

## Redundancy

There is redundancy in most speech and writing. When we say 'Tom is arriving by train at 4 p.m. tomorrow', the meaning is clear. It is still clear in the telegraphic form 'Tom arriving train 4 p.m. tomorrow'. Abbreviate to 'Tom arriving train 4 tomorrow' and ambiguity, uncertainty, creeps in. Is it 4 a.m. or 4 p.m.? This is no problem if the recipient knows there is a train at 4 p.m. but not at 4 a.m. If the recipient knew this and also that Tom *always* arrived by train, the necessary information could be communicated by 'Tom arriving 4 tomorrow'. If it were known (before the message were transmitted) that Tom was arriving at an unspecified time by train next afternoon, the essential information could be abbreviated to 'Tom arriving 4'.

Redundancy in everyday speech is, in part, a protection against distortion of messages in transmission by what is called in the jargon of pattern recognition *noise*. We know the problems of understanding a message on a crackly telephone line and of poor radio reception. A rumour is often a distortion or misinterpretation of a message during relay; the final version may lose its original meaning after transmission by a long chain of people or through a number of telecommunication booster or relay stations. I like that classic, I suspect apocryphal, story of the pre-decimal currency era about the army officer who ordered a member of his Company to whisper a message to his neighbour in the ranks, he in turn to whisper it to the next man, and so on. The message started as 'Send me reinforcements, I am going to advance' but returned as 'Lend me three and fourpence, I am going to a dance'.

## Distorted signals

Distortion of signals is a common difficulty when information is transmitted mechanically or electronically, and also in computer operations or photography. Studies of pattern distortions due to noise or loss of information form a major part of the subjects *information theory* and *communications theory*.

The science of *pattern recognition* deals with situations ranging from those where distortion, if any, is well understood and leaves little or no uncertainty, to cases involving considerable uncertainty. Fingerprints are widely used in detection because the odds are astronomically high against their being identical for two people. Recognition is here essentially devoid of uncertainty – it is simply a matter of checking records for a 'match'. In this enlightened age only very amateur criminals leave their fingerprints at the scene of a crime. Science has responded by looking for other near foolproof identifiers. Interest currently centres on unique biochemical patterns in, for example, saliva and in so-called *genetic fingerprinting*, especially useful in sex killings or rape cases where specimens of blood or semen can be used to deduce an attacker's virtually unique DNA characteristics.

Many simpler human features are more or less unique. We recognize friends by their build, appearance and perhaps by the clothes they wear, but such identifications are far from certain. We all know the embarrassed stranger who apologizes after heartily slapping one on the back 'Oh, sorry, I thought you were Joe'. With identical twins, relying on our senses of sight and sound for recognition may be hazardous if it is important to pick the right one.

### Modern research

Two well-researched areas of pattern recognition are:

 (i) 'teaching' computers to recognize symbols (writing or magnetic codes) or sounds (speech);
(ii) disentangling the information in, for example, pictures taken by satellites or displays like those on radar screens, from distortion and noise, and also combining information from several such sources to give a more complete picture.

The familiar parodies of numbers on the bottom of modern cheques – cheque number, account number and the bank sort code – are there to be read by electronic scanners. Lasers interpret the now familiar bar codes on packaged goods on the supermarket shelf. Automatic cash dispensers read information encoded on magnetic strips on plastic cards.

Progress is being made in getting machines to read ordinary handwriting or to recognize sounds. Here, the problems are formidable: those of us who are bad writers (the author included) know how easy it is to mistake a written 3 for a 5 or a badly written 5 for an 8. Similarly, a machine recognition of spoken commands faces formidable problems with dialect and other voice variations. Computer diagnosis, met in Chapter 12, is in essence a type of pattern recognition.

### Dirty pictures

We look at some simple applications falling in the second category above to illustrate some current ideas in pattern recognition. The technique is sometimes referred to as *cleaning up dirty pictures*; it can also be used to highlight precise features of general pictures.

The methods are based on the way pictures, from a satellite, say, are encoded for transmission. Each picture is divided into small elements called 'pixels' – a corrupted abbreviation of *picture elements*. For a black and white picture, sets of predetermined numbers of digits (usually binary digits, i.e. an ordered string of zeros and ones) are used to indicate the degree of darkness of each pixel. A string of zeros represents pure white; a string of ones, pure black. Shades of grey are represented by different combinations of zeros and ones in a string. The simplest case is where each pixel is either *black* represented by 1 or *white* represented by 0, and there are no intermediate 'greys'. From such simple beginnings, digital image transmission has developed to the stage where, for example, the Landsat satellite takes four simultaneous pictures each covering different parts of the light spectrum. Pixel information is transmitted digitally and can then be combined to give just one picture useful for determining, among other things, land use.

A complication is that the pictures are distorted by atmospheric effects and technical limitations of photography. Further, these distortions are different for light of different wavelengths. As well as distortions, there are blips, not unlike film grain marks or other

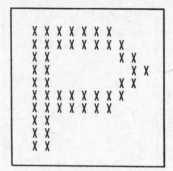

*Figure 14.1.* Digital image letter P.

faults, that show up in enlargements of an ordinary photograph. This is the pure noise component. In combining photographs to study distribution of forests or crops, the digital signals are *filtered* to reduce noise and eliminate some distortions. In addition, special transformations may be needed before several pictures are superimposed; these eliminate remaining differential distortions. The transformations use mathematical techniques akin to regarding the photographic plates as sheets of rubber which can be individually stretched until they match one another. The mathematics is formidable and the calculations required can only be done by large-capacity computers.

## Filters

We illustrate the basics of filtering out noise by a very simple example where each pixel is transmitted as a single digit, either 0 or 1. In practice a square centimetre of photograph might be divided into more than 60 000 squares, each represented as a pixel and transmitted as 0 or 1. Typical distortion may result in the edge of a black object on a white background becoming blurred, or a straight line becoming wavy. We suppose in our simple illustrations that *noise*, such as spots, shows as a black pixel where there should be a white one. Figure 14.1 shows a true representation of the letter P digitized as black (represented here by a cross ( × ) corresponding to a digit 1) on a white background (represented

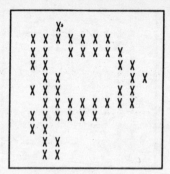

*Figure 14.2.* Letter P distorted.

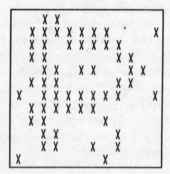

*Figure 14.3.* Letter P with distortion and noise.

by a blank corresponding to a digit 0). The square array consists of $12 \times 12 = 144$ pixels, shown as either a cross or a blank at the centre of each square representing a pixel.

Figure 14.2 is a distorted version of this picture. The number of black pixels is the same, but some have shifted position slightly. This is typical of distortion that may occur due to camera limitations, or when photographing from a satellite through a distorting medium, such as the atmosphere. Atmospheric distortion of the moon or sun when near the horizon, making them look bigger than they do when high in the sky, is a familiar example of this phenomenon.

```
        N
    N   S   N
        N
```

*Figure 14.4.* Pixel S and neighbours N.

In Figure 14.3 we have superimposed a simple form of noise on Figure 14.2. The noise is such that certain white pixels are replaced randomly by black ones; this corresponds to black blemishes on an 'all white' blank photographic plate. In this trivial example we do not allow the opposite type of noise, that of some true black pixels registering as white, although in many practical situations both forms of noise occur.

Figure 14.3 is a so-called *dirty picture*. Statistical methods of filtering to clean it use the principle that true pattern is represented by concentrations of pixels of one kind, while noise results (in our case) in more or less isolated black pixels. In practice, filters are usually highly sophisticated mathematical functions. The pixels are compared one at a time with many others in the neighbourhood. An isolated pixel very different from its neighbours (in our simple case black when neighbours are predominantly white) suggests noise.

To illustrate the principle of cleaning up we consider an over-simplified filter that operates by examining each cross (black pixel) and its neighbours above and below and to left and right forming a cross pattern. This is illustrated in Figure 14.4 for one pixel. S is the pixel of interest; the N's are its neighbours.

Suppose there is a cross in position S. Our filtering rule is that we accept this as real if at least two neighbours are also crosses; if it has none or only one neighbouring cross, we filter it out, regarding it as noise.

If we sweep through Figure 14.3, replacing crosses that do not have at least two neighbouring crosses by circles (O), we get Figure 14.5.

Check that each circle in Figure 14.5 has at most one neighbouring cross, and that each cross in Figure 14.5 has at least two neighbouring crosses, in Figure 14.3.

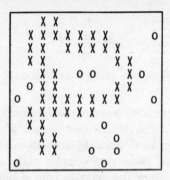

*Figure 14.5.* Pixels labelled ○ are removed by filter.

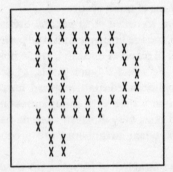

*Figure 14.6.* Effect of filter on Figure 14.3.

Deleting the circles, we get Figure 14.6; this is very like Figure 14.2. Removal of distortion now requires transformation techniques analogous to stretching a rubber sheet, or perhaps others that depend on prior knowledge about possible edge shapes that can arise in the true situation. We omit details here.

Finally, we note that a noise filter should have little effect on the true image we are trying to reproduce. However, it may also have only a minor effect on distortion like that in Figure 14.2. In Figures 14.7 and 14.8 we show the effect of filtering applied to Figures 14.1 and 14.2 directly. This shows that our filter has little effect when there are only certain forms of distortion and no random noise.

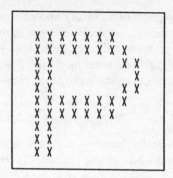

*Figure 14.7.* Filter applied to Figure 14.1.

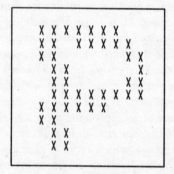

*Figure 14.8.* Filter applied to Figure 14.2.

## Another kind of filter

Filters have other uses. They may help detect subtle changes in near-identical photographs that have a lot of detail. For example, satellite pictures can be used to detect land planted with particular crops. If pictures taken a year apart are compared, it is possible to filter out everything except fields where different crops are grown in each year. These fields are then highlighted without the cluttering detail of the vast areas growing the same crops or otherwise unchanged. The same thing may be done in time-lapse photography, where an object is photographed many times at fixed intervals to see what changes occur with time. In medicine, for example, a photograph of an artery and surrounding tissue may be taken at one second intervals to study movement of a blood component. The pictures will be cluttered with unchanging detail in the artery wall and surrounding tissue. If this can be 'filtered' out, it is much easier to see how the blood component is moving.

### The lone walker

To illustrate the basics, suppose we have a series of 20 × 20 arrays of black and white pixels. The black pixels might represent people standing or walking in a city square. If the pictures are taken at regular time intervals and most people remain still, they will occupy the same position in successive frames taken, say, at three second intervals. Those who are walking will change position in a fairly regular manner. If the square is fairly crowded, it will be hard to pick these out by looking at successive frames. If we could *remove* the fixed people from all frames and superimpose the frames in correct order, we are left with the paths taken by those who are walking. Figure 14.9 represents six successive frames where each cross represents a person and all but one of these stands still throughout. It is tedious to compare frames pairwise visually to see which cross is changing position. However, if we superimpose frame (b) on frame (a) and use a filter that removes all crosses that coincide on both frames we are left with one cross (in different positions) on each frame which indicates the path of the person walking. We may then do the same for frames (b) and (c), (c) and (d), and so on.

Superimposing frames (a) and (b) in Figure 14.9, we would find all crosses coincide except for one cross which occurs in row 12 column 14 in (a) in row 11 column 15 in (b). Comparing (b) and (c), we find the odd cross is, as before, in row 11 column 15 in (b) and has moved to row 10 column 16 in (c). Finally, we find the position of this odd cross in the other frames are row 9 column 17 in (d), row 8 column 18 in (e) and row 7 column 18 in (f).

Superimposing all six frames and eliminating crosses common to all, we are left with the path of the one walker, shown in Figure 14.10.

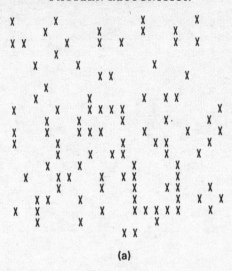

(a)

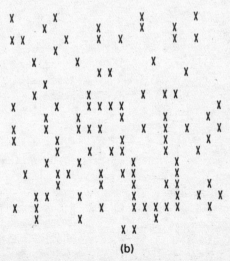

(b)

*Figure 14.9.* Time-lapse photographs of people in a square, one walking.

(c)

(d)

(e)

(f)

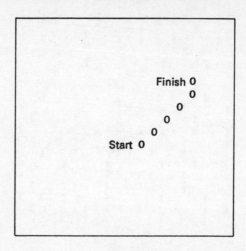

*Figure 14.10.* Path of person walking, all those standing still filtered out.

There was no noise in this example. We could use the same filter to detect several walkers and obtain their paths. Techniques like this have been extended to filter out cars travelling on a motorway at exactly some given speed (e.g. the authorized maximum speed limit) from those travelling either above or below that speed.

A wealth of detail about *digital image processing*, as filtering and transformation methods for clarifying pictures is called, is given in simple terms by Billingsley (1972).

## Some social consequences of pattern recognition

Modern technology has given us many new techniques for recognizing patterns. The social consequences are likely to prove important, with possibilities ranging from unwelcome Orwellian scenarios to very real benefits for at-present underprivileged groups.

Machines that can recognize speech or read the written word may eliminate a need for copy-typists or word-processor operators. Will this be a boon because it does away with tedious jobs, or a disaster that will throw many people out of work? Or will there be a spin-off to create even more jobs in the new technologies themselves? If so, will those displaced want, or be able to cope with, the more skilled jobs, or will they bemoan a loss of work that is emotionally undemanding?

## Ante-natal screening

Pattern recognition makes possible genetic screening at the early stages of pregnancy to determine if an unborn child is likely to be the victim of some genetically transmitted disease (p. 211) or subnormal as a result of a genetic deficiency. If the result is positive, the parents may then decide if the pregnancy should be terminated. Some people favour genetic screening as a routine during pregnancy; however, it shows up, not only genetic deficiencies, but also the sex of the foetus. If the results of the screening were made known to mothers (and the current social trend favours the patient being allowed to have medical informaton of this kind on request), there might be an increase in abortions on grounds like parents not wanting a child of a particular sex. After a few years, this could result in a large excess of, say, male over female births. If so, who can predict the consequences of the frustration of adolescent boys who have little chance of finding a partner? Would it be the end of family life as we know it? Might social unrest or crime bred of frustration become widespread?

## Automatons at war

Pattern recognition is also well suited to electronic warfare. Some ten years ago Batchelor (1978) envisaged the possibility of booby trap bombs that explode when they hear an individual speak or have a conversation in a certain language. He even suggested the possibility of totally mechanized war, fought by robot-like devices with little or no regard for their own survival. Is *star wars* a logical development of this idea?

### *The brighter side*

But all is not grim. Pattern recognition is a vital tool in mass screening for diseases. Machines to carry out computer diagnoses may be mass produced and exported worldwide to relieve pressures on overworked doctors in developing countries.

Satellite photographs that show world crop distributions may be invaluable at the humanitarian level for predicting famines and at the economic level for planning a better distribution of food and better management in agriculture.

The methods are likely also to help in detecting dangers inherent with what is probably one of the greatest threats to the world today – atmospheric pollution.

### A review

There is probably a consensus view that most applications of pattern recognition to date are beneficial – the use of machine-recognizable codes, the cleaning up of dirty pictures, the filtering out of unwanted photographic detail, and so on.

However, the peep into the future towards the end of this chapter reveals that there lurks a potential for misuse of, or evil application of, standard mathematical and statistical techniques for 'recognition' purposes. This is a good illustration of the way powerful techniques developed with the best of intentions may well be open to abuse in the wrong hands.

That such abuses may take place is an added uncertainty that goes with most scientific progress. In a very different field, genetic engineering, there are endless possibilities for good or evil, most of these probably undreamt of by the research workers who made these manipulations of the hereditary system possible.

Pattern recognition is just one field where vigilance is needed to stop abuses: vigilance by scientists, industrialists, politicians (at national and international level) and, not least, from the celebrated man in the street.

# EPILOGUE

We have described a mixed bag of successes and failures in dealing with uncertainty. Stories of success in meeting its challenges in medicine and in industry, for example, make quite cheerful reading. Slower progress in dealing with natural and man-made hazards, and our growing ability to produce more of the latter as technology advances, is a cause for concern.

What becomes clear from the success stories in Part II – Salk, improvements in industrial efficiency and quality of goods, benefits stemming from some aspects of pattern recognition, the steps we now take to check new drugs – is that, given enough relevant information, we can cope quite successfully with known and quantifiable uncertainties using the methods of classical probability theory and statistics, applying these with a blend of common sense and technical know-how.

From our failures – dealing with atmospheric pollution, with natural disasters, inability to cope with fresh dangers from new technologies – we learn that there are risks and hazards that can only be minimized by collaborative efforts. Such efforts have been hampered all too often by a lack of relevant information or by the misuse or misinterpretation of facts to suit interested parties.

It may seem cynical to say that, in the twentieth century, science is too important to be left to the scientists, industry too important to be left to industrialists, the environment too important to be left to environmentalists, health too important to be left to the medical profession, politics too important to be left to the politicians – but it is true.

The examples in this book, and there are dozens of others, tell why. Environmental pollution is an excellent illustration. No one group can make important decisions that are in any sense the best possible without the cooperation of one or more of the other groups.

The environmentalist who shuts his eyes to the impossibility of zero pollution is no more a realist than the industrialist who does not give a damn how much he pollutes the atmosphere. There are a few extremists in each camp, but most environmentalists and most industrialists take less divergent stances. Yet there remains a yawning gap between what the former think is necessary and what the latter are prepared to do about pollution.

So we need some umpires, indeed several kinds of umpire. Scientists to produce facts. Politicians and legislators to set out rules to reduce pollution and to protect the environment. Between these, we may need another group of experts: economists to cost various options and even psychologists or sociologists to consider social and emotive reactions to any proposals.

To make such umpiring systems work, scientists will need to become better than most are now at presenting and interpreting facts both to the opposing interests, environmentalists and industrialists, and to the other umpires, the politicians, legislators, economists, sociologists and psychologists.

Politicians will have to grab the nettle of uncertainty: forget the 'let's wait until we are sure' syndrome. Difficulties of international cooperation will have to be faced, not used as an excuse for delaying action. Refreshing signs of new political realism stem from the remarks by William Waldegrave quoted on p. 141. Let us hope he and his fellow politicians show equal enthusiasm in translating these views to actions.

The importance of psychological reactions to risk cannot be ignored, but the sociologist and psychologist have a duty to ensure that emotion is tempered by reason rather than fuelled by illogical fear. This aim will be achieved if the layman can be given a better appreciation of uncertainty.

If this book has shown something of the all-pervasive nature of uncertainty, demonstrated that we can cope with it rationally in some circumstances and shown the need for a more enlightened approach to its more frightening manifestations, it will have served its purpose. It gives no easy solutions, for there are none.

Franklin may have overstated and over-simplified his proposition. But he is right, we live in an uncertain world, one in which it is increasingly important to balance emotion with reason. It is time for all parties to look at all the relevant facts in their attempts, not to eliminate uncertainty, for that is impossible, but alleviate the consequences of its more diabolical manifestations.

# BIBLIOGRAPHY

## Further reading

There are many elementary statistical textbooks. One that expands upon many of the ideas in Chapters 4–6 at a leisurely pace is *Statistics Without Tears* by Derek Rowntree (Penguin, 1981). At a more detailed mathematical level *Elementary Statistical Methods* by G. B. Wetherill (Chapman and Hall, 1982) and *Statistics for Technology* by C. Chatfield (Chapman and Hall, 1983) are very sound texts on general statistics. The book by P. G. Moore included in the references gives a general account of a quantitative treatment of risk with emphasis on business and industrial situations. Environmental risks are covered in books by Fred Pearce and James Bellini also included in the references.

The three reports listed under Anon. also contain a wealth of interesting material.

Three books that deal with case histories of statistical applications are:

*Statistics: A Guide to the Unknown*, ed, J. Tanur (Holden-Day, 1972).
*Statistics and Public Policy*, ed. W. B. Fairley and F. Mosteller (Addison-Wesley, 1977).
*The Fascination of Statistics*, ed. R. J. Brook, G. C. Arnold, T. H. Hassard and R. M. Pringle (Marcel Dekker, 1986).

## References in text

Aitchison, J., and Kay, J. W. (1973). A diagnostic competition. *Bulletin of the Institute of Mathematics and its Applications* 9, 382–83.

Anon. (1983). *Risk Assessment. Report of a Royal Society Study Group.* London: The Royal Society.

Anon. (1985). *Report of the Acid Rain Enquiry.* Edinburgh: Scottish Wildlife Trust.

Anon. (1987). *Special Research Report No. 20 – Acid Rain.* London: Central Electricity Generating Board.

Barnett, V. D. (1982). *Comparative Statistical Inference*, 2nd edn. Chichester: Wiley.

Barry, J. (1968). General and comparative study of the psychokinetic effect on a fungus culture. *Journal of Parapsychology* 32, 237–43.

Batchelor, B. G. (1978). Some aspects of pattern recognition. In: *Pattern recognition. Ideas in practice*, ed. B. G. Batchelor, pp. 463–71. London: Plenum.

Bellini, J. (1986). *High Tech Holocaust*. Newton Abbot: David and Charles.

Billingsley, F. C. (1972). Digital image processing for information extraction. In: *Machine Perception of Patterns and Pictures*, pp. 337–62. London: Institute of Physics.

Bissel, A. F., and Pridmore, W. A. (1981). The UK average quality system and its statistical implications. *Journal of the Royal Statistical Society, A* 144, 389–418.

Blumenson, L. E. (1976). When is screening effective in reducing the death rate? *Mathematical Biosciences* 30, 273–303.

Bradshaw, A. (1985). Opening Address, *The Acid Rain Inquiry*. Edinburgh: Scottish Wildlife Trust.

Brown, J. (1987). Risk perception: expert opinion versus public understanding. *Economic and Social Research Council Newsletter* 59, 17–19.

Brownlee, K. A. (1955). Statistics of the 1954 polio vaccine trials. *Journal of the American Statistical Association* 50, 1005–13.

Burch, P. R. J. (1978). Smoking and lung cancer: the problem of inferring cause. *Journal of the Royal Statistical Society, A* 141, 437–77.

Doll, R., and Hill, A. B. (1964). Mortality in relation to smoking: ten years' observations on British doctors. *British Medical Journal* 4(1), 1399–1410, 1460–67.

Douglas, M. D., Elsworth, C. M., and Galbally, I. E. (1986). Ozone in near surface air. In: *Baseline 83–84*, ed. R. J. Francey and B. W. Forgan, pp. 41–42. Australia: Department of Science.

Downton, F. (1982). Legal probability and statistics. *Journal of the Royal Statistical Society, A* **145**, 395–402.

Fairley, W. B., and Mosteller, F. (1974). A conversation about Collins. *University of Chicago Law Review* **41**, 242–53. [Reprinted in *Statistics and Public Policy*, ed. W. B. Fairley and F. Mosteller. Reading, MA: Addison-Wesley (1977).]

Fischoff, B., Slovic, P., and Lichtenstein, S. (1982). Lay foibles and expert fables in judgements about risk. *The American Statistician* **36**, 240–55.

Francis, T., Jr (1955). An evaluation of the 1954 poliomyelitis vaccine trials – summary report. *American Journal of Public Health* **45**, 1–63.

Goldberg, J. D., and Wittes, J. I. (1981). The evaluation of medical screening procedures. *The American Statistician* **35**, 4–10.

Green, C. H., and Brown, R. A. (1978). Counting lives. *Journal of Occupational Accidents* **2**, 55–70.

Green, D. R. (1984). Talking of probability . . . *The Bulletin of the Institute of Mathematics and its Applications* **20**, 145–49.

Green, D. R. (1987). Probability concepts: putting research into practice. *Teaching Statistics* **9**, 8–14.

Harvey, A. C., and Durbin, J. (1986). The effects of seat belt legislation in British road casualties: a case study in structural time series modelling. *Journal of the Royal Statistical Society, A* **149**, 187–227.

Kennedy, I. R. (1986). *Acid Soil and Acid Rain.* Letchworth: Research Studies Press.

Lichtenstein, S., Slovic, P., Fischoff, B., Layman, M., and Combs, B. (1978). Judged frequency of lethal events. *Journal of Experimental Psychology: Human Learning and Memory* **4**, 551–78.

Meier, P. (1972). The biggest public health experiment ever. In: *Statistics: A Guide to the Unknown*, ed. J. Tanur, pp. 2–13. San Francisco: Holden-Day.

Molina, M. J., and Rowlands, F. S. (1974). Stratospheric sink for chlorofluoro-methanes: chlorine atom catalysed destruction of ozone. *Nature* **249**, 810.

Moore, P. G. (1983). *The Business of Risk.* Cambridge: C.U.P.

Napley, D. (1982). Lawyers and statisticians. *Journal of the Royal Statistical Society, A* **145**, 422–26.

Neave, H. R. (1981). *Elementary Statistical Tables.* London: Allen and Unwin.

Newell, D. (1982). The role of the statistician as an expert witness. *Journal of the Royal Statistical Society, A* **145**, 403–9.

Pearce, F. (1987). *Acid Rain.* London: Penguin.

Peto, R. (1975). Smoking and bronchitis. *Bulletin of the Institute of Mathematics and its Applications* **11**, 62–63.

Reid, D. D. (1972). Does inheritance matter in disease? The use of twin studies in medical research. In: *Statistics: A Guide to the Unknown*, ed. J. Tanur, pp. 77–83. San Francisco: Holden-Day.

Sprent, J. I. (1987). *The Ecology of the Nitrogen Cycle*. Cambridge: C.U.P.

Sprent, P. (1988). *Understanding Data*. London: Penguin.

Tiao, G. C. (1983). Use of statistical methods in the analysis of environmental data. *The American Statistician* **37**, 459–70.

Tversky, A., and Kahneman, D. (1974). Judgement under uncertainty: heuristics and biases. *Science* **185**, 1124–31.

Walker, M. A. (1987). Interpreting race and crime statistics. *Journal of the Royal Statistical Society*, A **150**, 39–56.

Watson, G. S. (1982). Stratospheric ozone. Observations and data analysis. *The American Statistician* **36**, 312–16.

# INDEX